SOLID-STATE BATTERIES VIA ADDITIVE TECHNIQUES.

Jennifer C. Anderson

Advances in additive manufacturing (AM) processes are continuously opening up the material design space, providing scientists with opportunities to explore the relationship between structure, processing, and materials properties in new contexts. The first project presented in this book presents the design and refinement of a novel, polyborane-based solid electrolyte, whose design and investigation were motivated by the advent of additively manufactured, 3D electrodes, which could play a pivotal role in enabling next-generation batteries that can store more energy without sacrificing power. The first iteration of this electrolyte was synthesized by hydroborating polybutadiene with 9-borabicyclo(3.3.1)nonane (9-BBN). The resultant poly(9-BBN) was then reacted with precise amounts of n-butyllithium (n-BuLi), an organolithium reagent, to create the final polymer electrolyte. The polymer electrolyte films were assembled into a custom apparatus for impedance measurements, and though found to be ionically conductive, these measurements were not consistent, even within films made from the same batch of polymer in solution.

This necessitated the modification of the electrolyte into a UV-cured version, which was achieved by hydroboration of polybutadiene using 9-BBN. The resulting poly(9BBN)-co-polybutadiene is treated with lithium tert-butoxide (LiOtBu) and crosslinked to produce a precursor resin, which is then drop cast onto PTFE spacers, UV-cured for 5 minutes, dried, and assembled into coin cells for electrochemical impedance spectroscopy (EIS) and into pans for differential scanning calorimetry (DSC). The ionic conductivity (σ) of the PBEs as measured by EIS as a function of molar salt ratio, $r = mol_{Li}/mol_B$, does not track with their measured glass

transition temperatures, T_g or the activation energies, E_a, extracted from fitting the Vogel-Tammann-Fulcher (VTF) equation to the conductivity data. Beyond $r = 0.33$, values for E_a and T_g demonstrate insensitivity to increasing concentration, while σ continues to change with concentration and reaches a maximum at $r = 0.75$. Morevoer, measurement of ionic conductivity of control PBE films without boron on the polybutadiene backbone confirms that the presence of Lewis-acidic boron groups is necessary for ionic solvation and conduction. Further analysis that compared the PBEs to a well-studied PEO-based electrolyte in the literature[1] through the calculation of a reduced conductivity, σ_r, to co ntrol fo r po lymer vi scosity an d se gmental motion, revealed that PBEs obtain optimal conductivity at higher salt concentrations than PEO, and that their ionic conductivities are far below that of PEO. We posit that we are observing a mechanism of ionic conduction in a glassy regime partially decoupled from the relaxation of the polymer host. We attribute these effects to the strong interaction between the Lewis-acidic boron centers and the strongly Lewis-basic tert-butoxide anions, which limits ionic conductivity by suppressing motion of the anions and presenting a large activation barrier for motion of Li+, which is optimized at high concentrations where the distance between the boron-anion centers is sufficiently small to increase the probability of a hopping event from one center to another.Nanorods fashioned from noble metals are ideal for maximizing extinction of electromagnetic radiation, which is necessary for plasmonically active materials in numerous applications, from contrast agents for biological imaging to effective obscurants. Key challenges that prevent nanorods from being employed for these technological applications include the prohibitively expensive cost of Au and Ag, their lack of requisite thermal and chemical stability, and the limitations in resolution and attainable feature sizes produced by existing wet chemistry techniques. The second project in this book focuses on the development of an AM process to create arrays of TiN-coated microbridges with lengths of 4.749 ± 0.048   m, cross-sections with dimensions of 0.692 ± 0.015   m by 2.256 ± 0.077   m, and effective aspect ratios of 3.368, that are capable of attenuating light reflected from a TiN-coated sapphire substrate by more than 80% in the mid-infrared (mid-IR), as measured by Fourier Transform Infrared (FTIR) spectroscopy. FTIR spectroscopy measurements further reveal attenuation of light transmitted through the same TiN-coated structures by up to 35% in the near- to mid-IR. These results indicate a promising pathway for AM of plasmonically active microparticles with broad reflectance and transmittance attenuation of light in the near- and mid-IR.

TABLE OF CONTENTS

Abstract   v

Published Content and Contributions   vii

Table of Contents   vii

List of Illustrations   x

List of Tables   xii

Chapter I: Introduction   1

1.1 A brief history of materials science and additive manufacturing . .1

1.2 Why additive manufacturing matters for energy storage, and why

    electrolytes are the roadblock   3

1.3 Why additive manufacturing matters for the design of optically

active materials   6

    1.4 book overview   7

Chapter II: Models of ionic conduction in polymer electrolytes . . . .

.   9

2.1 Theory of ionic conduction in linear polymer electrolyte systems   .

.   9

2.2 Empirical treatment of ionic conduction in polymer systems . . . . .

               15

2.3 Measuring ionic conduction in polymer electrolytes: Electrochemi-

cal impedance spectroscopy   23

2.4 Thermal characterization of polymer electrolyte systems: Differen-

    tial scanning calorimetry   26

    2.5 Conclusion   28

Chapter III: Polyborane-based, ether-oxygen-free electrolytes for solid-state,

lithium and lithium-ion batteries   30

3.1 Challenges of solid polymer electrolytes: Tight coupling between

ionic conduction and segmental polymer motion 30

3.2 Fabrication of electrolyte films and measurement of their ionic con-

ductivity 32

3.3 Results 38

3.4 Discussion 38

3.5 Summary 42

Chapter IV: Exploration of ionic conduction in polyborane-based polymer

electrolytes for lithium and lithium-ion batteries 45

4.1 Introduction 45

4.2 Experimental methods 47

4.3 Results 49

4.4 Discussion 54

4.5 Conclusion and future directions 65

ChapterV:AdditivemanufacturingforIR-responsive,TiN-coatedmicrostruc-

tures 69

ix

5.1 Design of plasmonic materials and how additive manufacturing can expand the design space 69

5.2 Fabrication of TiN-coated microbridges 70

5.3 Material characterization 72

5.4 Optical characterization 72

5.5 Optical response 75

5.6 Discussion 76

5.7 Summary and outlook 80

Chapter VI: Conclusion 81

6.1 Summary and future research directions 81

6.2 Outlook 82

Chapter1

INTRODUCTION

1.1 A brief history of materials science and additive manufacturing It has been said that the history of humankind is the history of materials. So pivotal to human development were the discovery and mastery of certain materials that eras in human history carry their names: Stone, bronze, and iron. With the advent of each age, humans were able to master new abilities and organize into more complex societies. Understanding how to work stone into tools and weapons allowed our hunter-gatherer ancestors to become more efficient resource gatherers, and enabled the development of sedentary societies built around animal husbandry and agriculture. With the arrival of metal-working, namely the ability to mold and process metal ores into bronze and iron, ancient humans were able to make more durable and reliable metal iterations of tools like plows that increased agricultural productivity, and forge weapons and armor to wage war, conquer, and establish administrative control over ever-larger expanses of land. More interestingly, these advancements happened in parallel in disparate parts of the world, a testament to early human observation and experimentation [2]. The ability to work bronze and later iron arose independently in regions like the Indus Valley, Mesopotamia, Europe, Egypt, and ancient China [2]. Peoples who lived along the Pacific coast of what is now Peru first became master gold- and silver-smiths, making a variety of elaborate pieces for ceremonial and decorative purposes, and eventually mastered copper alloys as well [2–4].

Though ancient peoples mastered the craft of metallurgy, their understanding of the relationshipbetweenprocessingandpropertieswasprimarilyempirical[2]. Lacking the means to probe the structure and dynamics of the materials they processed at the atomistic level materials scientists do today, explanations for why certain materials possessed certain properties, or why different processing techniques resulted in different material outcomes, often bordered on the fantastical or cosmological. Ancient metallurgists did not have a sophisticated understanding of processes like diffusion or kinetics, for instance, and as a result attributed the strength of steel blades to the liquids used in the quenching process, or other unusual practices [5]. One ancient method called for quenching hot steel with the urine of a red-haired

boy; another, more gruesome technique called for plunging a cooling blade into the body of a muscular slave, whose strength would then be imparted to the sword [5]. It was not until the 19th century and the dawn of the industrial age that more advanced theory arose to explain the empirically observed relationships between material structure, processing, and properties, giving birth to the field we know today as materials science [2].

The technological advances of the 19th century, in addition to a more formal and theoretically solid framework for understanding material properties, allowed for the production of a variety of materials with specific properties at great scales. Today we produce a plethora of products, from steel to plastics, in great quantities and a variety of shapes and sizes. Common to most forms of modern manufacturing is that they are subtractive, meaning that the desired geometry of a product is achieved by removing material [6, 7]. Examples of subtractive processes include milling, machining, laser cutting, and some lithography processes [6, 7]. Many subtractive manufacturing processes result in a large amount of material waste, and do not allow for rapid prototyping, the design of 3D materials with fine features, or custom part design [7]. In contrast to subtractive manufacturing, additive manufacturing (AM) implies the creation of objects using constituent materials from the ground up, either piece-by-piece or layer-by-layer, until the final product is achieved [6, 7]. The advent of modern AM began with the patent of the first vat photopolymerization device in the 1980's by Kodama [8], which functioned by using UV light to cure a vat of polymerizableresinlayerbylayertoachieve3Dstructures. Inthetimesince,additive manufacturing methods have become more sophisticated, and include variations of Kodama's original design like vat stereolithography, which uses patterned UV light to print customizable 3D structures layer by layer, and two-photon lithography (TPL) direct laser writing (DLW), which uses a finely defined laser-point to cure

3D structures out of photoresins with feature sizes down to hundreds of nm (Figure 1.1). The advent of these technologies has allowed for the expansion of the materials design space, enabling the fabrication of architected materials that exhibit novel and tunable properties not possible in their bulk counterparts. This has positive implications for a variety of applications, from enabling next-generation 3D battery technologies that could accelerate the transition to electric vehicles and renewable energy generation, to tailor-made nano- and microparticles that respond to light in the infrared. Chapter 1.2 provides an overview of the promise of 3D batteries, and why the design of a capable, solid polymer electrolyte is crucial in realizing their potential. Chapter 1.3 briefly introduces the field of plasmonically active nano-

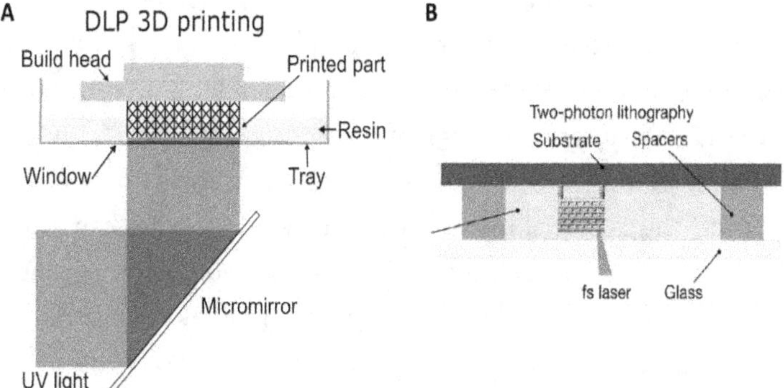

Figure 1.1: Demonstration of two AM processes: Vat stereolithography and twophoton lithography (A) A schematic of a typical vat stereolithography setup. UV light is modulated and directed using a mirror through a window on the bottom of a vat with photopolymerizable resin, onto which is projected a pattern to print a layer of a computer designed structure. The build head pulls the cured layers upwards, allowing for the rest of the structure to print layer by layer. Figure adapted by permission from John Wiley & Sons - Books: Wiley - VCH Verlag GmbH & Co. KgaA, Advanced Energy Materials, reference [9], Copyright 2021. (B) Typical set up for two-photon lithography, direct laser writing. Curable resin is placed directly onto a functionalized substrate, which is then bounded by two Kapton tape

spacers   andcappedwithaglasssslide. Alaseristhenfocusedintoanellipticalvoxel, whichis guided through the use of moving mirrors to selectively cure the resin according to a computer designed structure. Figure adapted with permission from [10]. Copyright 2020, American Chemical Society.

and microparticles, and discusses how AM can allow for the tunable design of such particles with target properties in the infrared. Chapter 1.4 presents these problems in the context of the battery and optical work presented in this book, where the unifying theme is AM expanding the space of material properties for meaningful scientific investigation.

1.2 Why additive manufacturing matters for energy storage, and why electrolytes are the roadblock

The importance of lithium-ion batteries to modern life cannot be understated. They are a critical pillar in our electronic and information age,   powering   all   manner   of portabledevicesfromsmartphonestolaptops.   Iflithium-ionbatteriesaretodrivethe   greenrevolution,   however, theywillneedtostoremoreenergythantheydocurrently,   without sacrificing power and at the very least maintaining their physical footprint. This is crucial to ensuring that electric vehicles can achieve ranges sufficient to compete with gas-powered cars [11], and to enabling large-scale renewable energy storage that can manage the rapid fluctuations of our modern electrical grids [12]. The current lithium-ion battery consists of three major parts: A graphitic anode, a porous polymer separator soaked in electrolyte, and a metal-oxide cathode, as shown in Figure 1.2. To store more energy in this configuration requires more Li atoms, which in turn requires more C atoms in the anode and more metal and O atoms in the cathode [13]. This translates to thicker electrodes, as the material mass needed to house the Li atoms increases, which increases the diffusion distance for lithium ions into the electrode, negatively impacting the lithium-ion battery's power density [14, 15].

There are two potential solutions to circumventing this energy/power tradeoff: the first involves simply replacing the graphitic anode with lithium metal, which can increase energy density without requiring a thicker carbon anode, as shown in the bottom of Figure 1.2. This solution is precluded, however, by the current class of electrolytes, which typically consist of lithium salts dissolved in organic solvents that do little to suppress the formation of dendrites and can result in a catastrophic short circuit, which leads to combustion of the electrolyte [13, 16, 17]. The second solution, which aligns more closely with the focus of our research group and the theme of this book, involves the creation of 3D-architected electrodes, like the ones shown in the bottom of Figure 1.3. The loss of power density associated with thicker electrodes can be circumvented if the thickness is used as a space to carve out architected stuctures, like the carbon anode posts demonstrated in Figure 1.3. The cathode can be conformally coated around the anode posts, and as the battery is made thicker, more energy can be stored while maintaining the same distance between and into the electrodes, thus avoiding power losses.

Key to realizing either of these solutions is a solid electrolyte that is ionically conductive, mechanically resilient, and in the case of 3D-batteries, capable of being conformally coated onto complex 3D geometries. Natural candidates for this role are polymers like poly(ethylene oxide) (PEO), which have been known to dissolve lithium salts and conduct lithium ions since the 1970's [18, 19], and can be cured in situ when needed [20]. PEO and its analogues, however, suffer from low ionic conducitivity and cationic transference number, which precludes their use as electrolytes in 3D batteries. This problem motivates the projects described in chapters 3 and 4, which look at polymer alternatives that seek to invert the mechanism of solvation and conduction in PEO by replacing Lewis-basic ether oxygen atoms

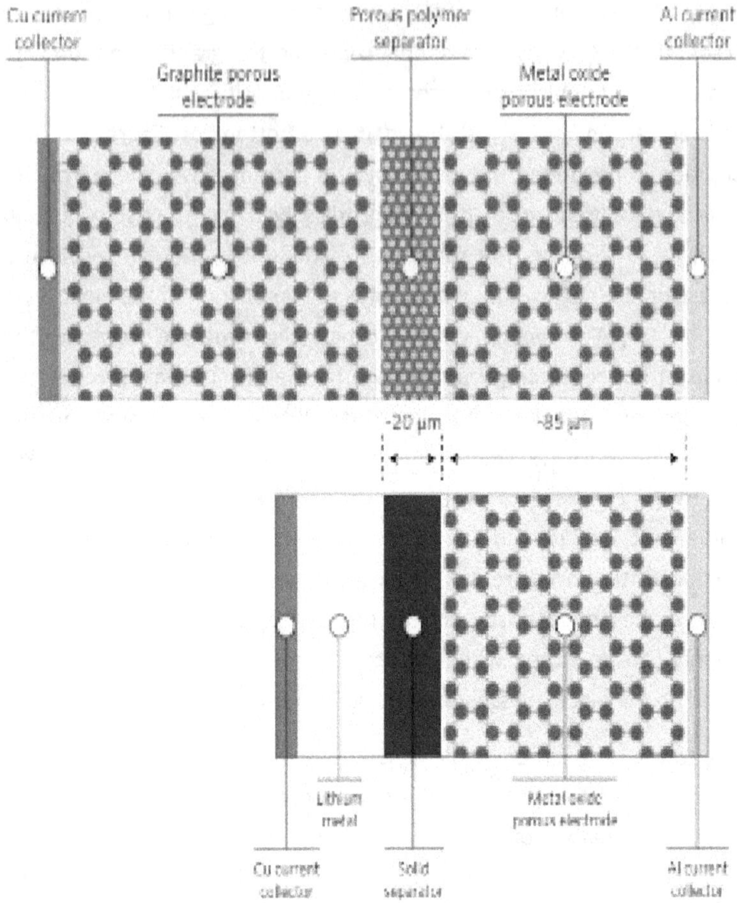

Figure 1.2: Diagram of a typical lithium-ion battery on the top, and the configuration ofaproposedlithiummetalbatteryonthebottom. Storingmoreenergyinthetypical lithium ion configuration requires thicker electrodes, which comes at the cost of decreased power. Lithium metal anodes could circumvent this constraint. Figure adapted by permission from Springer Nature Customer Service Centre GmbH: Springer Nature, Nature Energy, reference [13], Copyright 2018.

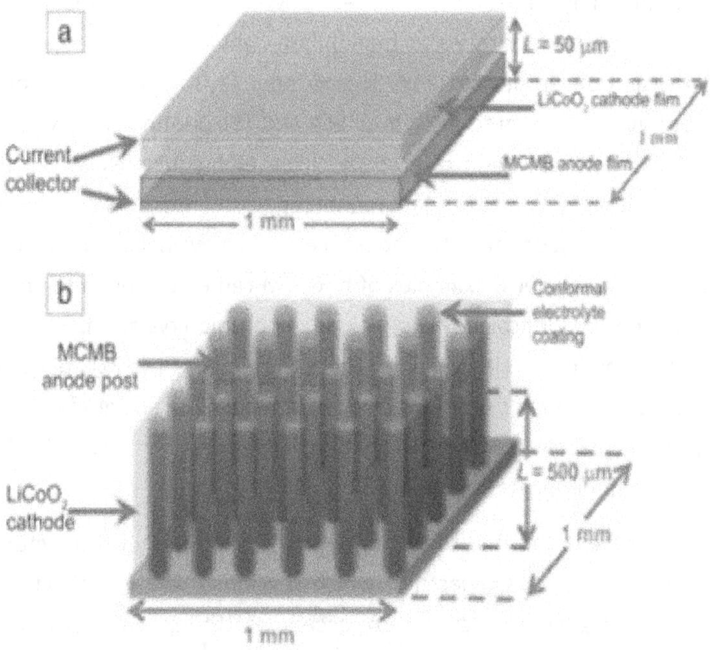

Figure 1.3: Mock-up illustrating the advantages of 3D-architected batteries. Architected electrodes, like the ones shown in the bottom, exploit the third dimension of height to allow for more energy to be stored by making the battery thicker, while ensuring that diffusion distances between and into the electrodes remain constant, resulting in no power loss. Figure adapted by permission from Springer Nature Customer Service Centre GmbH: Springer Nature, MRS Bulletin, reference [14], Copyright 2011.

with Lewis-acidic boron atoms. In seeking to understand the mechanism of ionic conduction in these novel polymers, we hope to unlock some insight into making polymer electrolytes that can meet the challenge of enabling architected, additively manufactured batteries.

1.3 Why additive manufacturing matters for the design of optically active

materials

The benefits incurred by advancements in the realm of AM extend as well to the design of optical materials with tunable properties, finely controlled feature sizes, and complex geometries. Given the customizability afforded by modern AM processes, all manner of devices can be fabricated to manipulate light at a variety of scales. Classical optical devices like lenses and sensors can be rapidly prototyped and printed using both SLA and TPL printing methods [21], and the fine resolution offered by these techniques allows for the introduction of architected features that wouldotherwisebedifficultorimpossibletoachieveusingtraditionalmanu facturing techniques. Examples of how AM can be exploited to create new kinds of optical devices include: Photonic crystals made from titania woodpiles with tunable band gaps in the IR [22]; photonic crystals that exhibit negative refraction in the IR [23]; and graded-index materials that can function as Luneberg lenses [24], flat lenses [25, 26], and field-tapering devices [27].

It should be noted that the aforementioned additively manufactured devices were all dielectric, which excludes the possibility of manipulating free electrons to access scattering and absorption phenomena like plasmonic resonances. Plasmons are nonradiative oscillations of free electrons induced by externally applied electromagnetic fields [28, 29]. Nanoparticles made from conductive materials like Au or Ag exhibit plasmon resonances in the visible and near-IR, and by changing their lengths or aspect ratios, their plasmon resonances can be tuned. There is great interest of pushing the plasmon resonances of nano- and mircroparticles further into the IR, thus enabling technologies that improve absorption in solar panels [30], allow direct detection of analytes in biological samples [31], or function as obscurants in the near- and mid-IR [32, 33]. AM offers tight control over and reproducibility of microparticlesizeanddimensions, advantageousindesigningparticleswithtargeted plasmon resonances. Chapter 5 discusses our efforts to employ AM to push the design of plasmonically active microparticles into the IR, which we accomplish by printing structures using a TPL DLW process, followed by coating these structures in TiN via atomic layer deposition (ALD).

1.4 book overview

The aim of this book is to explore the properties of novel materials whose design is driven either directly or indirectly by AM. Chapter 2 provides a primer on theoretical and empirical models of ionic conduction in polymer electrolytes, in addition to a brief survey of experimental techniques used for their characterization: electrochemical impedance spectroscopy (EIS) and differential scanning calorimetry (DSC). Chapter 3 describes the first pass at making an ionically conductive polyborane-based polymer electrolyte where ionic conduction is mediated through solvation of anions, the challenges encountered, and how these challenges informed a redesign of the chosen electrolyte system. Chapter 4 probes the mechanism of ionic conduction in the redesigned, UV-cured polyborane-based polymer electrolytes, using the experimental techniques and the empirical models of ionic conduction described in chapter 2. This modified electrolyte demonstrated more reliably consistent ionic conductivity as a function of concentration, exhibited unique thermal behavior, and by virtue of being curable can be employed to coat architected electrode structures. Examining how ionic conductivity and glass transition temperature evolve as a function of concentration suggests that a new mechanism of ionic conduction may be at play.

Chapter 5 discusses the AM of TiN-coated microbridges via TPL DLW and ALD. Characterization of the materials' optical response reveals broad attenuation of reflectance and transmittance in the near- and mid-IR, and reflectance measurements reveal a large attenuation at about 10 μm, a possible indicator of a plasmonic resonance.

Chapter 6 summarizes the findings in each of these projects, discusses avenues of future research, and mentions potential large-scale implications for the work presented in this book.

Chapter 2

MODELS OF IONIC CONDUCTION IN POLYMER ELECTROLYTES

2.1 Theory of ionic conduction in linear polymer electrolyte systems

Polymer electrolyte systems can vary in complexity, but at their most basic level consist of linear polymer chains dissolving salts to provide mobile ionic species. Assuming a strong electrolyte with complete dissolution of salt by the polymer chains of the melt, and that the polymer electrolyte itself is well above the glass transition temperature, ionic conduction can be understood as intrinsically coupled to the diffusivity of individual polymer chains. More precisely, the mechanism of this polymer chain diffusion changes depending on the molecular weight of the polymer. Below the molecular weight at which polymer chains become entangled, M_{ent}, chains in a polymer melt can be understood in a manner analogous to polymer chains in a good solvent at dilute concentrations. In this dilute regime, chains stretch out and behave close to ideally, and are well-approximated by spherical blobs which do not overlap and are characterized by their radius of gyration, R_g (See Figure 2.1). Diffusion of polymers on length scales on the order of R_g or greater in this unentangled concentration regime are characterized by diffusion of these spherical blobs, and this mechanism is described by the Rouse model of diffusivity, which we proceed to describe here according to the treatment in Rubinstein and Colby. [34]

In approximating an unentangled, linear polymer system, the Rouse model assumes that a polymer chain can be represented as N beads connected by springs of rootmean-square size, b, and interact only with each other through the connecting springs. Each bead is assumed to have its own friction coefficient, ζ, and the total friction coefficient of the entire Rouse chain is then

$$\zeta_R = N\zeta. \tag{2.1}$$

The Einstein relation, which relates the diffusion coefficient, D, of a particle, and its friction coefficient, ζ, is:

$$D = \frac{kT}{\zeta}. \tag{2.2}$$

Figure 2.1: Low molecular weight polymer melts behave similarly to polymers in good solvent at dilute concentrations. The polymers are defined by spherical blobs with radius of gyration, R_g, and these blobs do not overlap.

To obtain the diffusion coefficient of the Rouse chain, we then substitute equation 2.1 into 2.2 to obtain:

$$D_R = \frac{kT}{\zeta} = \frac{kT}{N\zeta}. \tag{2.3}$$

R

Having derived the diffusion coefficient for the Rouse chain, we now return to our model of polymer chains in solution. Below the entanglement molecular weight, M_{ent}, in the polymer melt, diffusion of entire polymer chains is akin to the diffusion of spherical polymer blobs of radius R_g in a good solvent at dilute concentrations. What happens when the molecular weight of the polymer exceeds M_{ent}? This is described by the scenario where the total volume of the spheres in solution is on the same order of the total solution volume, V:

$$nR_g^3 \sim V, \tag{2.4}$$

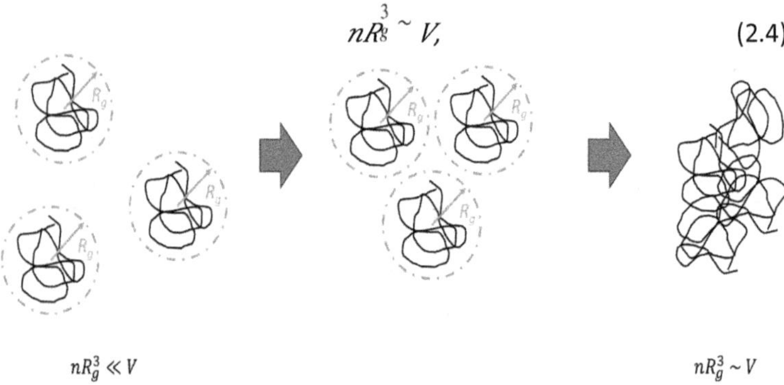

Figure 2.2: As the concentration of the polymer increases, the spherical blobs defined by R_g become more tightly packed until they begin to overlap, which occurs when $nR_g^3 \sim V$. At this point the polymer solution is said to have entered the semidilute regime. This approximates the behavior of polymer melts beyond the entanglement molecular weight, M_{ent}.

where n is the number of polymer chains in the solution. At this point, the chains overlap and become indistinguishable from one another, and we are said to be in the semidilute (entangled) regime (Figure 2.2).

Here, the diffusion of the chains on the order of their length or greater is no longer correlated with the diffusion of spherical blobs of radius R_g, but now confined by the presence of other, surrounding chains. Describing chain diffusivity here requires a different theory, and this problem was tackled by Pierre-Gilles de Gennes, who approached it by imagining a single polymer chain sliding like a snake confined in a tube

(Figure 2.3) in a motion he described as reptation. In this model, we consider the diffusion of the entangled polymer chain on the order of the length of the confining tube, which we take as the average contour length, $\langle L \rangle$. In order to arrive at an expression for $\langle L \rangle$ in terms of the chain molecular weight, we begin by considering the width of the tube. The confining tube can be approximated by a constraining harmonic potential, where the minima trace out the primitive path along the center of the tube (Figure 2.3). The monomers in the chain are constrained by this potential, but thermal fluctuations on the order of kT are allowed. These fluctuations define the width of the confining tube, a, which can be understood as the end-to-end distance of an entanglement strand of N_e monomers:

$$a \approx b N_e^{1/2}. \tag{2.5}$$

The tube can then be thought of as consisting of an average of N/N_e sections, each

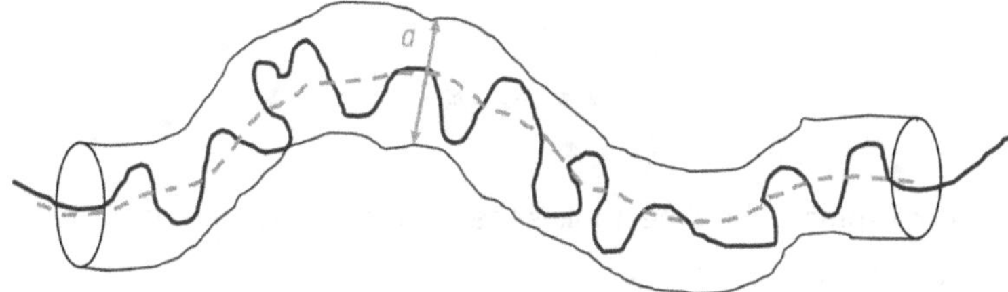

Figure 2.3: The confinement of an entangled polymer chain can be approximated by a tube. The width of the tube, a, is defined by a harmonic confining potential, and corresponds to fluctuations of the polymer chain on the order of kT away from the potential minima, which lie along the dotted line through the center of the tube that defines the primitive path.

with size a, and the chain itself a random walk of of these entanglement strands:

$$N^{1/2}$$

$$R \approx a \, \frac{}{N_e} \, .$$ (2.6)

We can then approximate the average contour length of the primitive path in terms of monomer size and molecular weight of the tube by taking the product of the entanglement strand size, a, with the average number of entanglement strands:

$$\langle L \rangle \approx a \, \overline{\frac{}{N_e}} \approx \overline{\frac{}{N_e^{1/2}}} N \qquad\qquad bN$$ (2.7)

With the the chain length and average contour length in hand, we can approximate relaxation time for entangled polymer chain reptation.

Like the Rouse model for unentangled dynamics, we again assume that friction coefficient for the chain is the sum of the individual friction coefficients for the

monomers, $\zeta_r = N\zeta$, and using the Einstein relation:

$$D_{ent} = \frac{\_\_kT}{N\zeta},$$ (2.8)

we can estimate the time it takes the chain to diffuse out of a tube of average length $\langle L \rangle$ as:

$$\tau_{rep} \approx \frac{\langle L \rangle^2}{D_{ent}} \approx \frac{\zeta b^2 N^3}{kTN_e} = \frac{\zeta b^2 N^3}{ktN_e}.$$ (2.9)

The first part of the relation in equation 2.9 is simply the Rouse time of an entanglement strand containing N_e monomers:

$$\tau_e = \frac{\zeta b^2}{kTN_e},$$ (2.10)

And whose diffusion would be characterized as:

$$D_e = \frac{kT}{N_e \zeta}. \tag{2.11}$$

Both τ_e and D_e are constant, independent of the molecular weight of the chain. As a brute approximation, the reptation relaxation time is proportional to the cube of the molecular weight, which is not far off from experimental observations, where the scaling exponent is found to be 3.4.

Having derived expressions for the diffusion coefficient of polymer chains in the
Rouse (Equation 2.3) regime and whole chain relaxation time in the entangled (Equation 2.10) regime, we can now make connections to how this diffusive motion is related to cationic mobility in polymer electrolytes, following the work of Shi and Vincent.[35] We begin with the following assumption: When in the Rouse regime below the entanglement molecular weight of the polymer, ionic conduction owed to polymer motion can be understood to have two components, one related to segmental motion of the chains, that is, for polymer chain relaxations on the order of entanglement strand relaxation, τ_e; and the second owed to Rouse diffusion of entire polymer chains along with solvated cations. We can then write the overall diffusion coefficient for ionic motion in the Rouse regime, D^+ as follows:
$R,total$

$$^+{}_{R,total} D^+{}_e D^+ \tag{2.12} D = + R,$$

where D^+ is the diffusion of ions due to relaxations of a single entanglement strand,
e

and D^+ is the diffusion of ions due to whole chain Rouse diffusion. Equation 2.12
R

can in turn be used in conjunction with a Nernst-Einstein formulation to make an estimate of cationic conductivity owed to polymer motions in the Rouse regime:

$$\sigma_{R,total} = D_{R,total} \frac{nq^2}{kT} + \quad + \tag{2.13}$$

$$\sigma_{R,total} + \frac{nq^2}{kT} = D_{+e} \quad D_{+R} \quad + \tag{2.14}$$

As the molecular weight of the polymer increases beyond its entanglement threshold, whole chain diffusion will cease, and the only movements available to the chains would be segmental motion of entanglement strands on the order of time τ_e and reptation relaxation of whole polymer chains on the order of time τ_{rep}. Cationic diffusion in this regime is assumed to happen on timescales $\tau \ll \tau_{rep}$ and much closer to $\tau \approx \tau_e$, so measured ionic conductivity beyond the entanglement molecular weight will plateau at values driven by entanglement strand relaxations. The term associated with D^+ in equation 2.14 can then be treated as a constant and lower e limit, which we denote σ^+_e:

$$\sigma_{R,total} = \sigma_e + {} + \frac{nq_2}{kT} D_R. + \tag{2.15}$$

Taking the expression for D_R in equation 2.3 and plugging it into 2.15, we arrive at:

$$+ {} + nq_2\,kT \tag{2.16}$$

$$\sigma_{R,total} = \sigma_e +$$

$$\sigma_{R,total} = \frac{\overline{k_B T} \; \overline{N_7}}{\frac{2}{N_7}} \quad \sigma_e + \tag{2.17}$$

$$\sigma_{R,total} = \sigma + \frac{e}{N} , \tag{2.18}$$

Where the constant $A = nq^2/\zeta$. Equation 2.18 shows that as molecular weight of the polymer increases, the second term becomes negligible and the ionic conductivity is a constant determined primarily by the process associated with the relaxation of individual entanglement strands, which is expected beyond the polymer's entanglement threshold. Shi and Vincent developed this model, but observed trends in steady state current and diffusion coefficient measurements in polyethylene oxide (PEO) electrolytes as proxies for ionic conductivity.[35] Semilogarithmic plots of this data against molecular weight showed a decrease in steady state current or diffusion until the entanglement molecular weight was reached, at which point the measurements plateaued and remained insensitive to further increases in molecular weight (See Figure 2.4). Teran et al. applied this model directly to measurements of ionic conductivity of PEO-LiTFSI electrolytes of varying molecular weight and found that the trends held (See Figure 2.5).

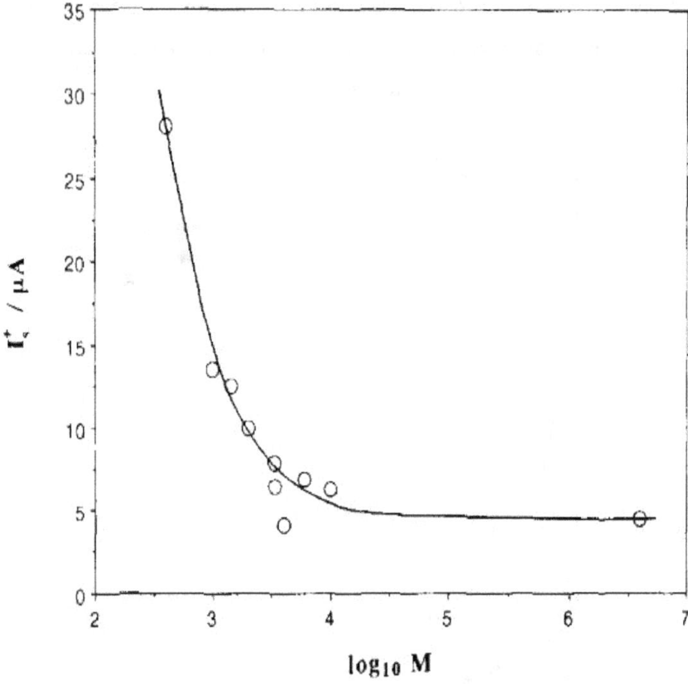

Figure 2.4: Figure from Shi and Vincent [35] demonstrating constant steady state current beyond the entanglement threshold of a PEO-based electrolyte with $LiCF_3SO_3$ as a salt. Reprinted from [35] with permission from Elsevier.

2.2 Empirical treatment of ionic conduction in polymer systems

The empirical study of polymer electrolytes is not limited, however, to the study of the effect of molecular weight alone on ionic conductivity. The studies of Shi and Vincent [35] and Teran et al. [36] fixed temperature and salt concentration, which themselves can have important effects on ionic conductivity. Salt concentration can notably affect the mobility of polymer chains by introducing transient crosslinks between them, decreasing the degree of configurational entropy and elevating the temperature, T_g, at which the polymer transitions into a glassy regime where chain motionislimited. Otherfeatureslikechemicalcrosslinkingcansimilarlyeffectthese changes. Describing the effects of these factors on polymer properties like chain mobility and correlated, empirically measured polymer electrolyte ionic conducitivity will require a more comprehensive, phenomenological formulation. The approach developed from here forward pulls from Rubinstein and Colby,[34] Williams et

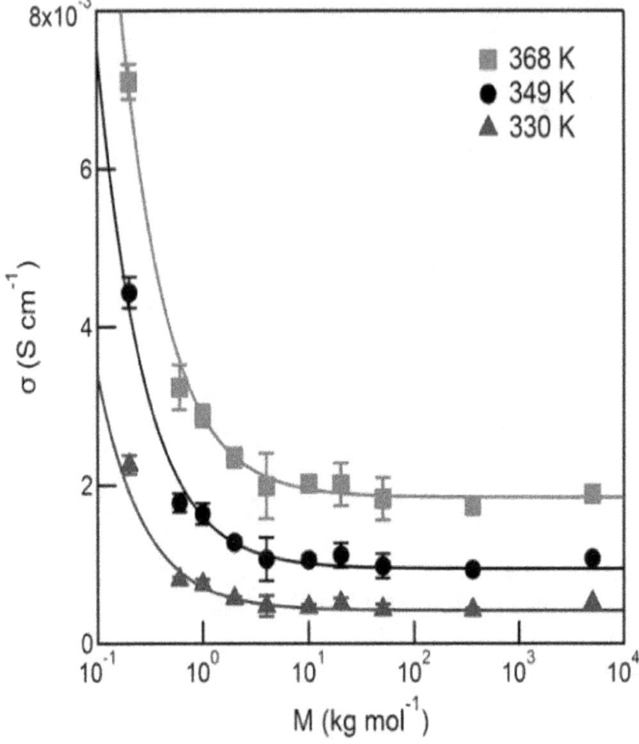

Figure 2.5: Figure from Teran et al. [36] demonstrating constant ionic conducitivity in a PEO-based electrolyte with LiTFSI as a salt beyond the entanglement threshold at three different temperatures. Reprinted from [36] with permission from Elsevier.

al.,[37] Ratner and Shriver,[38] and Peleg [39].

It is known that measurable polymer properties like modulus, relaxation times, and viscosity exhibit a dependence on temperature. Many models of polymer dynamics, including the Rouse and entangled models discussed in the previous section, experience changes in relaxation times across all modes as a function of temperature according to the following relationship:

$$\tau \sim \frac{\zeta}{T},\qquad(2.19)$$

where ζ is the friction coefficient. This understanding is intuitive, because as temperature rises, molecular motion becomes faster and corresponding relaxation times decrease, and when temperature decreases, motion slows and relaxation times increase. Similarly, properties like modulus are understood to behave according to:

$$G(t) \sim \rho T, \text{ where } \rho \text{ is}\qquad(2.20)$$

the polymer mass density; and viscosity according to:

$$\eta \approx \tau G(t) \sim \rho \zeta\qquad(2.21)$$

Because the relaxation times for all modes have the same temperature dependence, it should be possible to superimpose measurements of properties like relaxation time, modulus, and viscosity for the same polymer system taken at different temperatures onto each other to obtain a single master curve. This is the time-temperature superposition principle, and is typically exploited through the use of a multiplicative, unitless shift factor defined by the ratio of a measured property of interest at one temperature to the same property measured at an arbitrarily chosen reference temperature (in many cases chosen to be the glass transition temperature, T_g). One example would be the multiplicative shift factor, a_T, for relaxation time, defined as follows, using the relaxation time defined according to equation 2.19:

$$a_T = \frac{\tau}{\tau_0} = \frac{\zeta / T}{\zeta_0 / T_0}\qquad(2.22)$$

$$\frac{\zeta T}{\zeta_0}.\qquad(2.23)$$

$$= \zeta_0 T$$

By the same reasoning, we can define a shift factor for both the modulus, using equation 2.20:

$$b_T = \frac{G(\tau)}{G_0(\tau)} = \frac{\rho T}{\rho_0 T_0}; \qquad (2.24)$$

and for viscosity, using equation 2.21:

$$\frac{\eta}{\eta_0} = \frac{\rho \zeta}{\rho_0 \zeta_0} = a_T b_T; \qquad (2.25)$$

where the shift factor turns out to be a product of the shift factors for modulus and relaxation time. Application of these shift factors to measurement of modulus

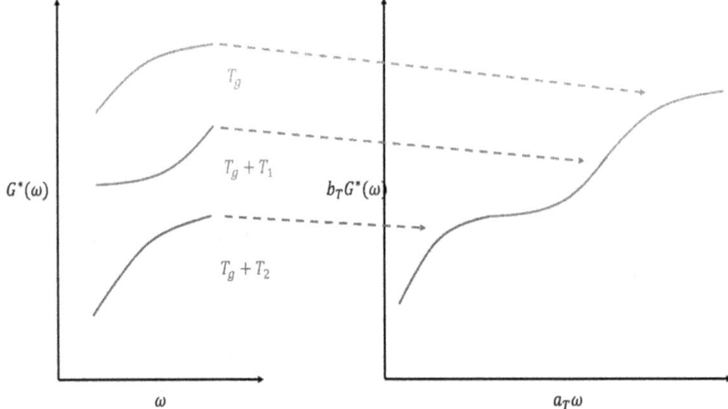

Figure 2.6: Illustration of time temperature superposition principle. Measurements are taken over the same frequency range, ω, at three different temperatures, with T_g as the reference. The use of the proper shift factors allows for the creation of a master curve over a wider frequency range.

relative to a chosen reference temperature, T_0, for example, would then take the following form:

$$G(t,T) = b_T G\left(\frac{t}{a_T}, T_0\right); \qquad (2.26)$$

for measurement of modulus in time, or:

$$G^*(\omega,T) = b_T G^*\left(\omega a_T, T_0\right); \qquad (2.27)$$

for measurment of complex modulus in frequency space. These equations provide experimentalists the ability to shift data taken at different temperatures over the same time or frequency range beyond what is typically possible in the lab, allowing for predictionofpolymerpropertiesbeyondexperimentallyobservablerange s, provided that the viscoelastic response observed is linear. This principle is illustrated in figure 2.6.

Not only do polymer properties like modulus, viscosity, and relaxation time depend on temperature, but the friction coefficient, on which viscosity and relaxation depend, also exhibits a temperature dependence that is not well-understood, and the shift factors used for time temperature superposition must also account for this. Therearevariousphenomenologicalassumptionsforapproximatingthete mperature dependence of the viscosity (and by extension friction), the most simple of which is the Arrhenius formulation, where:

$$\eta \sim exp\left(\frac{E_a}{kT}\right). \qquad (2.28)$$

The advantage of the Arrhenius formulation is that it captures the dependence of viscosity well at sufficiently high temperatures, where the activation energy, E_a, associated with relaxation, is constant. The Arrhenius model fails when polymers approach T_g, atwhichpointthedensity, andmoreimportantlythefrictioncoefficient, changedramaticallyduetocrowding. Closetothe T_g, thefrictioncoefficientchanges by roughly a factor of 10 when changing the temperature by 3K, and far above T_g

$(T > T_g + 100)$, the temperature would have to change by 25K to effect a similar change.

One approach to addressing this issue of temperature dependent dynamics of polymers was developed by Williams, Landel, and Ferry in their 1955 paper,[37] where they assumed that the crowding effect close to T_g in glass forming liquids (of which polymers are generally a subset) could be modeled by exploiting the concept of free volume. They assumed that the majority of the liquid volume, V, is occupied by molecules, and that the remaining free fraction, fV, is available to accommodate molecular motion. As the polymer passes T_g and is firmly in the glassy regime below T_g, the available free volume goes to zero. They began by modeling the behavior of viscosity on the availability of free volume according to the Doolittle equation:

$$\ln \eta \sim \ln A + B \frac{1 - f}{f},\tag{2.29}$$

where A and B are constants that can be fit to account for the properties of the polymer being observed, and the fraction $(1 - f)/f$ represents the ratio of occupied to free volume. Rewriting the above in exponential form, absorbing the $1 - f$ term into the constant, B, and assuming A on the order of unity we obtain:

$$\eta \sim \exp \frac{B}{f}.\tag{2.30}$$

To account for the change in free volume around T_g, we make another phenomenological assumption, namely that the free volume exhibits a linear dependence on temperature shifted according to an idealized glass transition temperature, T_∞:

$$f = \alpha_f (T - T_\infty).\tag{2.31}$$

The term T_∞ is known as the Vogel temperature and is typically observed 50K below the T_g of many polymer systems, where the free volume goes to zero. The motivation for using T_∞ instead of T_g follows from the fact that T_g typically observed usingtechniqueslikedifferentialscanningcalorimetryarevariable,dependingonthe rate of the applied temperature sweep to a sample of interest. Practically, this means that at some point sufficiently below the range of variability for T_g measurements, one can assume a polymer system to be completely in a glassy state. The prefactor α_f is the thermal expansion coefficient of free volume, which is taken as a constant, and has units of $1/T$.

We can now deploy equations 2.30 and 2.31 to calculate the appropriate shift factor for viscosity, with the assumption (often made in practice) that the shift factor b_T is of order unity:

$$a_T b_T \approx a_T = \frac{\eta}{\eta_0} = \frac{\exp\left(\frac{B}{\alpha_f(T-T_\infty)}\right)}{\exp\left(\frac{B}{\alpha_f(T_0-T_\infty)}\right)} \quad (2.32)$$

$$= \exp\left(\frac{B}{\alpha_f(T-T_\infty)} - \frac{B}{\alpha_f(T_0-T_\infty)}\right) \quad (2.33)$$

$$= \exp\left(\frac{B}{\alpha_f}\left(\frac{1}{T-T_\infty} - \frac{1}{T_0-T_\infty}\right)\right) \quad (2.34)$$

$$= \exp\left(\frac{B}{\alpha_f}\,\frac{T_0-T}{(T-T_\infty)(T_0-T_\infty)}\right) \quad (2.35)$$

$$= \exp\left(-\frac{B}{\alpha_f(T_0-T_\infty)}\,\frac{T-T_0}{T-T_\infty}\right) \quad (2.36)$$

$$= \exp\left(-\frac{B}{\alpha_f(T_0-T_\infty)}\,\frac{T-T_0}{T_0-T_\infty+T-T_0}\right) \quad (2.37)$$

Having arrived at equation 2.37, we can then recognize the following terms as constants:

$$C_1 = \frac{B}{} \quad (2.38)$$

$$\alpha_f(T_0 - T_\infty)$$

$$C_2 = T_0 - T_\infty, \tag{2.39}$$

And substituting them back into equation 2.37, we arrive at:

$$\frac{\eta}{\eta_0} = \exp \frac{-C_1(T - T_0)}{C_2 + T - T_0}, \tag{2.40}$$

Or in logarithmic form:

$$\ln \frac{\eta}{\eta_0} = \frac{-C_1(T - T_0)}{C_2 + T - T_0}. \tag{2.41}$$

Equation 2.41 is known as the Williams, Landel, Ferry, or WLF equation, and is used to obtain shift factors for viscosity measurements by fitting measured η/η_0 data plotted against $T - T_0$, with C_1 and C_2 as the fitting parameters. The WLF equation does a good job of capturing the dynamics of disordered polymer systems near and above their glass transition temperature, which is often chosen for the reference temperature, T_0. It must be noted, however, that because of the empirical and unsubstantiated assumptions made in the derivation above, the WLF equation is understood as a phenomenological and not theoretical treatment of the temperature dependence of polymer dynamics.

In addition to the Doolittle and Arrhenius formulations, one could also arrive at the WLF equation with the assumption that viscosity, η, behaves according to the formulation of Vogel, Tammann, and Fulcher,[40–42] which in a slightly modified form assumes that viscosity scales with temperature shifted by the Vogel temperature according to the following fashion:

$$\ln \eta \sim \ln A + \frac{\overline{}}{i\,(T - T_\infty)} \qquad B$$

$$(2.42) \qquad \underline{}$$

$$\eta \sim A exp. \qquad (2.43)$$

$$k(T - T_\infty)$$

A is a preexponential factor with units of viscosity, B is a factor with units of energy and serves as a pseudoactivation energy, and k is the Boltzmann constant. The VTF formulation above, as it is known, typically also carries a prefactor of $T^{1/2}$ in the exponential form, but this effect is negligible as the temperature approaches the vicinity of T_∞, and is left out of this analysis to simplify the derivation that follows. Unlike the Doolittle equation, the VTF equation accounts directly for the effect of the glass transition temperature on viscosity, as opposed to using the free volume approach. Using the same approach to derive the WLF equation from the Doolittle equation that begins in 2.32, we obtain:

$$a_T = \frac{\eta}{\eta_0} = exp\left[\frac{B}{k(T - T_\infty)} - \frac{B}{k(T_0 - T_\infty)}\right], \qquad (2.44)$$

Which has the same form as equation 2.33. We can then rewrite 2.44 as:

$$\frac{\eta}{\eta_0} = exp-\frac{B}{k(T_0 - T_\infty)}\frac{T - T_0}{T_0 - T_\infty + T - T_0}, \qquad (2.45)$$

Where we identify the following constants:

$$C_1 = \frac{B}{k(T_0 - T_\infty)} \qquad (2.46)$$

$$C_2 = T_0 - T_\infty, \qquad (2.47)$$

Which upon substitution into 2.45 results again in the WLF equation:

$$\ln \eta = \frac{-C_1\,(T - T_0)}{\eta_0\,C_2 + T - T_0} \tag{2.48}$$

Polymer data like viscosity can be fit with either the VTF or WLF equations, and the transformations in equations 2.46 and 2.47 allow the experimentalist to toggle between each formulation. This demonstration of the equivalence of the VTF and WLF equations is important in the context of polymer electrolytes, because the dynamics of the polymer, namely its viscosity, have been observed to affect the electrolyte's ionic conductivity.[38] Moreover, when plotting ionic conductivity on a logarithmic axis as a function of temperature within 100K of either T_g or the idealized T_∞, the observed trend is not linear, as would be expected for behaviors adequately described by the Arrhenius formulation, but curved, indicating that ionic conductivity, like viscosity, changes not only as a function of temperature, but as a function of how close the polymer electrolyte is to its glass transition temperature,[38] which makes intuitive sense given the cooperative nature between polymer motion and ionic conductivity, as explored in section 2.1. Given this observation, researchers likeArmand[43], whorecognizedthepotentialfor polymer salt complexes based on PEO to serve as battery electrolytes, posited formulations of the equivalent VTF and WLF equations for modeling the temperature dependence of ionic conductivity. Beginning with the assumption that ionic conductivity behaves according to the VTF form:

$$\sigma \sim A\exp \frac{B}{k\,(T - T_\infty)}, \tag{2.49}$$

And following the same procedure to derive equations 2.41 and 2.48, we arrive at the WLF formulation for ionic conductivity:

$$\ln \frac{\sigma}{\sigma_0} = \frac{C_1(T - T_0)}{C_2 + T - T_0} \qquad (2.50)$$

The constants C_1 and C_2 here are the same as those in equations 2.46 and 2.47. The power of extending the WLF and VTF formulas to describing ionic conductivity lies in the fact that changing the concentration of ionic species in polymer electrolytes results in changes to their T_g, and by extension their T_∞. Because polymer dynamics and ionic conductivity can vary greatly close to T_g and T_∞, the WLF and VTF equations can accommodate this behavior by allowing either C_1 and C_2 to vary (in the WLF form), or B and A to vary (in the VTF form), to fit ionic conductivity data, in addition to other polymer properties like viscosity. Moreover, with the appropriate choice of T_∞, measured ionic conductivity or viscosity data from polymer electrolytes of different concentrations can be collapsed onto a common reduced temperature axis, $T - T_\infty$. This allows the experimentalist to control for changes in T_g across polymer electrolytes that vary only in their concentration of ionic species, ensuring conductivity or viscosity of these systems is compared over an identical temperature range offset from the unique T_∞ at each concentration. This principle is demonstrated in depth in chapter 4.

2.3 Measuring ionic conduction in polymer electrolytes: Electrochemical impedance spectroscopy

Inordertoprobetheinterplaybetweenapolymerelectrolyte'sioniccondu ctivityand polymer properties like T_g, we first deploy electrochemical impedance spectroscopy (EIS) to measure its ionic conductivity. EIS probes the response of electrochemical systems by observing their response to an externally applied, AC voltage $V(\omega)$, typically on the order of tens of mV, at different frequencies, ω (Figure 2.7A). Elements of electrochemical systems behave much like combinations of electrical circuit elements such as resistors and capacitors; charging of capacitor-like elements can give rise to a measurable phase delay, δ between the applied $V(\omega)$ and the current response, $I(\omega)$, which varies as a function of ω (Figure 2.7A). Furthermore, by transforming Ohm's law into frequency space, we can generalize the concept of DC resistance, R, to AC impedance, Z, a complex-valued quantity defined by a magnitude, $|Z(\omega)| = V(\omega)/I(\omega)$; and an angle in the complex plane, δ (Figure 2.7B). By sweeping over a sufficiently large space of ω, typically from 1 MHz to 0.1 mHz for measuring ionic transport, different values of $|Z(\omega)|$ and δ can be obtained and a corresponding curve can be traced out in the complex plane. Simple RC circuits typically manifest as semicircles in complex-plane plots, as shown in Figure 2.7B, where R = 100 Ω and C = 0.1 μF. At the low frequency limit, $\omega \rightarrow$ 0, the impedance of the capacitor, C, defined as:

$$Z_C = \frac{1}{j\omega C}, \qquad (2.51)$$

where j is the imaginary unit, goes to infinity, and all of the current goes through R.

This corresponds to the intersection the semicircle with the Re[$Z(\omega)$] axis, which returns the DC resistance of the circuit, which is simply R, which in our example was chosen to be 100 Ω.

By analogy, the response of an electrochemical system under test can be modeled by an appropriate equivalent circuit model. Polymer electrolyte films sandwiched between two stainless steel spacers can be modeled with the circuit shown in the bottom of Figure 2.7B [44, 45]. As the electrolyte film is polarized by an applied $V(\omega)$, positive and negative charges amass at different ends of the electrolyte along the interfaces with the steel spacers, forming double-layer capacitances, modeled by C_d in the equivalent circuit. The bulk response of the electrolyte itself is modeled by a simple RC, or Voigt, element between the C_d, where R_b is the bulk ionic resistance of the electrolyte, and C_b the bulk capacitance of the entire spacer-electrolyte-spacer stack. This Voigt element is used to fit the dominant semicircular feature in the impedance spectra obtained from electrolyte films, which allows us to extract their

R_b.

It must be noted, however, that all of the capacitors in the equivalent circuit shown in Figure 2.7B are modeled as ideal. In practice, the phenomena in electrochemical systems assigned to capacitors are generally not purely capacitive; they can dissipate energy like resistors as well. This non-ideality in systems modeled with Voigt elements manifests in a depressed semiciricle in a complex-plane plot, and proper

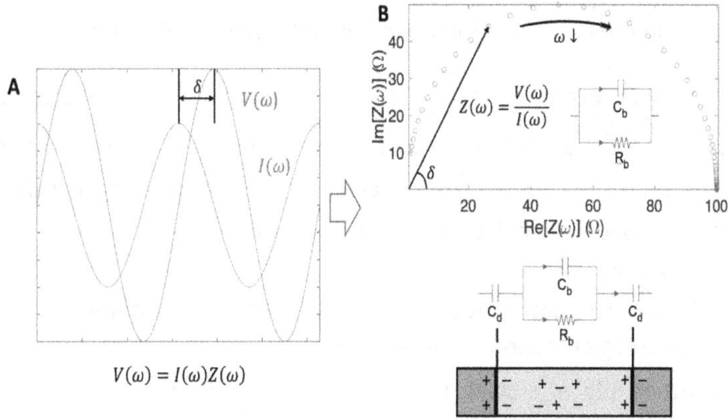

Figure 2.7: Primer on EIS. (A) Illustration of the basic form of the impedance experiment. An AC voltage, $V(\omega)$, is applied to some electrochemical system under test, and its current response, $I(\omega)$, is observed. Because many electrochemical systems exhibit both dissipative and capacitive qualities, there is a phase lag, δ, between the applied $V(\omega)$ and the observed $I(\omega)$. (B) The concept of DC resistance, R, is generalized to the complex-valued quantity, Z, by accounting for the phase lag, δ, in addition to the ratio $V(\omega)/I(\omega)$, which gives the magnitude of Z. By varying the frequency of the applied $V(\omega)$ over a sufficiently large range, different values for $|Z|$ and δ can be obtained, effectively parametrizing a curve in the complex plane known as an impedance spectrum. By fitting an impedance spectrum with an appropriate equivalent circuit model, meaningful information about the system under test, like bulk ionic resistance, R_b, can be extracted. The example at the top shows the semicircular spectrum associated with simple RC circuits; here R = 100 Ω and C = 0.1 μF. The model on the bottom shows the equivalent circuit model used to fit impedance spectra obtained

from a polymer electrolyte film (in yellow) sandwiched between blocking electrodes (gray, in this case the stainless steel spacers).

fitting of this feature requires replacing the capacitors in an equivalent circuit with a constant phase element, or CPE, which is defined as:

$$Z_{CPE} = \frac{1}{Q(j\omega)}\alpha, \tag{2.52}$$

where Q has units of capacitance and α is a parameter that goes from 0 to 1 and is determined by fitting.[44, 45] When $\alpha = 0$, the expression for Z_{CPE} simply reduces to a resistor, and when $\alpha = 1$, Z_{CPE} becomes an ideal capacitor. Once this correction is made, bulk ionic resistances, R_b can be reliably extracted from impedance spectra of solid electrolyte films and used to calculate their ionic conductivity.

2.4 Thermal characterization of polymer electrolyte systems: Differential scanning calorimetry

The other piece crucial for proper characterization of polymer electrolyte systems from the perspective of the WLF/VTF framework is their glass transition temperature, T_g. This can be obtained through the technique of differential scanning calorimetry, or DSC, where a sample of interest is heated and cooled at a constant ramp rate and the flow of heat in and out of the sample is measured. The process works by measuring up to 5g grams of the sample of interest into a pan, which is then placed onto one of two platforms within the measurement chamber of a DSC machine (Figure 2.8). The other platform holds an empty pan as a reference. Both platforms are on a stage that is heated uniformly at a constant rate, and thermocouples attached to each platform measure the temperature of each sample. The sample will experience temperatures different from that of the reference pan because the sample material will absorb or release heat at transitions, like those when it changes from a glassy to rubbery state, or when it melts. A computer reads this difference in observed temperatures between the samples, ΔT, through the thermocouples attached to each measurement platform, and produces a plot of heat flow (Q^*, with units of J/s) against temperature (T^*, with units of T/s), as shown in Figure 2.9.

The heat flow curve in figure 2.9 is a general example of the phenomena that can be observed upon heating a polymer sample. The glass transition, T_g, is characterized by the step-like feature, and the melting point, T_m, is demonstrated by the downward facing peak. Endothermic transitions result in more heat flow into the sample and face downwards in the graph, while exothermic processes (not pictured in figure 2.9) would face upwards. It must be noted that T_g is not technically a phase transition, and can vary depending on the rate of heating and cooling. Faster temperature

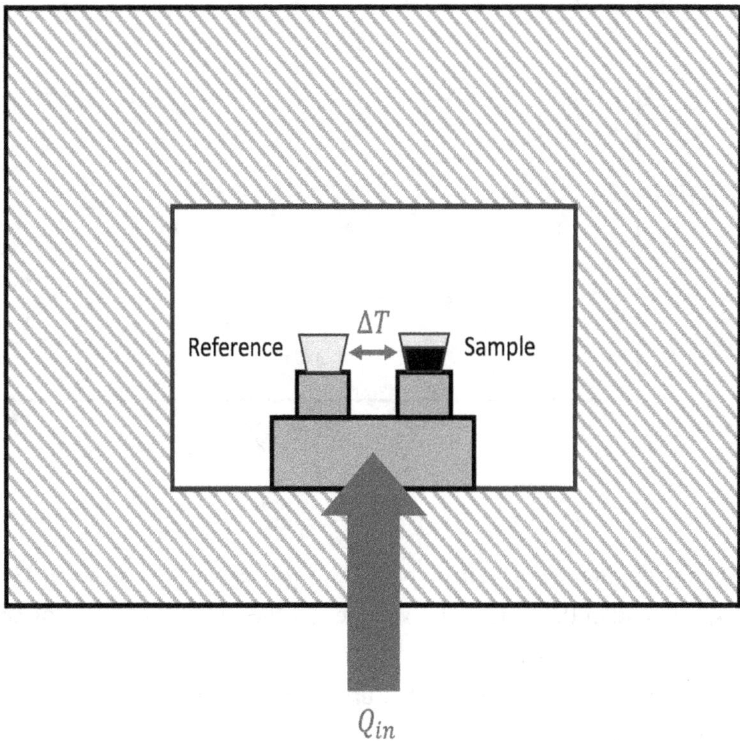

Figure 2.8: Simplified Demonstration of DSC experimental setup. Reference pan and sample pan are both heated uniformly, and thermocouples measure the temperature differences between both pans. Differences in observed temperature are used to calculate the heat flow into or out of the sample.

sweeps will generally produce a more pronounced glass transition feature in DSC experiments.

Extracting a value for T_g can be done in different ways, but one standard means of doing so is by calculating the derivative of the heat flow, Q^*, with respect to temperature, T^*, in the vicinity about a step-like feature in a heat flow vs temperature plot, as shown in Figure 2.10. The inflection point in the middle of the step manifests as a peak when dQ/dT is plotted against temperature, and this corresponds to the T_g of the sample. This technique is used to measure the glass transition from polymer electrolyte samples described in chapter 4.

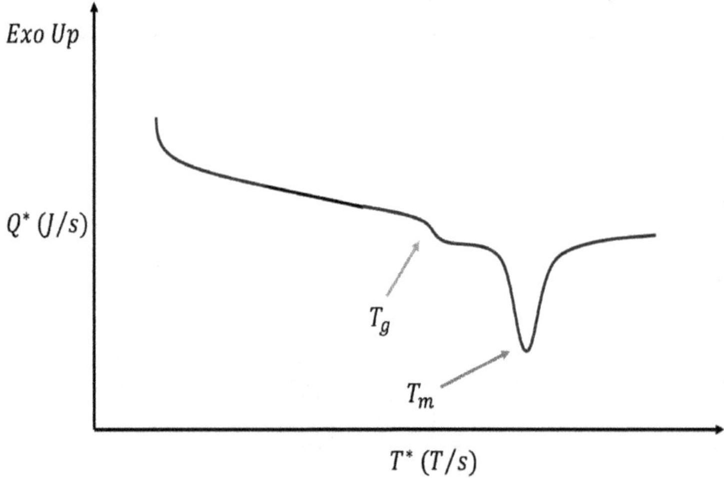

Figure 2.9: Sample graph of heat flow, Q^*, against T^*, from a DSC experiment. In this particular example, exothermic behavior is plotted upwards on the vertical axis.

T_g is visible in the step-like feature, and T_m is given by the large endothermic peak.

2.5 Conclusion

In this chapter, we developed the theory of ionic conduction for linear polymer electrolytes based on the Rouse and entangled models for polymer dynamics, while recognizing the limitations of this theory for describing how polymer ionic conducitivity could be affected by factors like changing concentration and crosslinking. In light of this challenge, we deployed the time-temperature superposition principle to develop the phenomenological models of the WLF and VTF equations to characterize properties like viscosity, which can capture effects from salt concentration and crosslinking. With additional assumptions we adapted these models to describe ionic conductivity of polymer electrolytes more generally, and will use the VTF model heavily in chapter 4 to analyze polyborane-based polymer electrolytes.

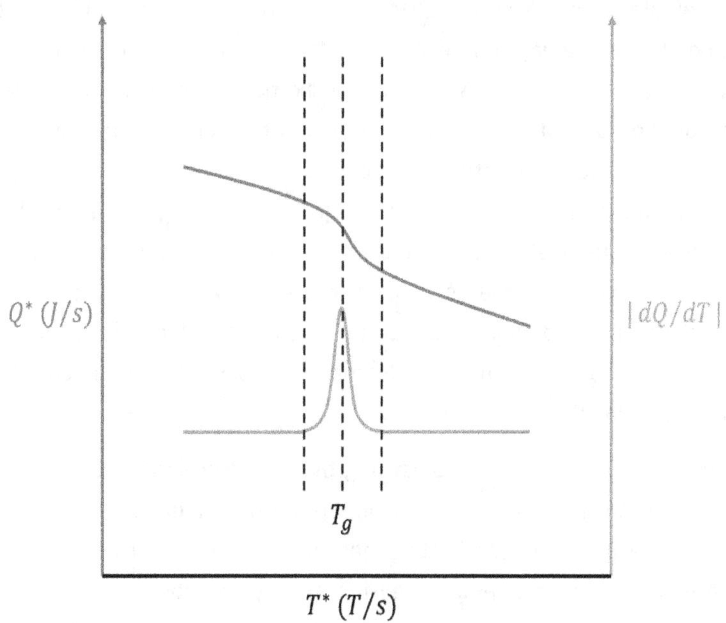

Figure 2.10: Extracting T_g from DSC heating curve using dQ/dT.

Chapter 3

POLYBORANE-BASED, ETHER-OXYGEN-FREE ELECTROLYTES FOR SOLID-STATE, LITHIUM AND LITHIUM-ION BATTERIES

This chapter's introduction has been adapted in part from:

F. J. Villafuerte et al. "Exploration of Ionic Conduction in Polyborane-based Polymer Electrolytes for Lithium and Lithium-ion Batteries (*In preparation*)".

3.1 Challenges of solid polymer electrolytes: Tight coupling between ionic
conduction and segmental polymer motion

The pressures of anthropogenic climate change necessitate the development of alternatives to fossil fuels. Solar and wind energy figure among the leading renewable energy options, but their use is hampered by their intermittency; peak hours for wind and solar production do not overlap with periods of high energy consumption, and large scale energy storage is needed to balance renewable energy production and consumption on the grid [11, 47]. Improved portable electrical energy storage is a crucial component in the sustainability equation: batteries that can store more energy and charge more quickly would be a boon in accelerating the transition to wind and solar, in addition to the adoption of electric vehicles, which is hampered by long charging times [11, 13].

Lithium-ion batteries are hamstrung by the intercalation compounds that form their electrodes, which limit the total amount of energy they can store [13]. Modern battery design makes use of thin films for battery components, and increasing energy density requires either increasing the areal footprint or making them thicker [13], the latter of which can increase diffusion distance for lithium ions into the electrode and negatively impact power density [14, 15]. One way to overcome this storage limitation involves swapping the graphitic anode for metal lithium, which can increase the energy density by more than 50% without sacrificing power density [13]. Stable cycling of a cell with a lithium metal anode is precluded by the formation of dendrites, which grow on the anode surface and eventually reach across the electrolyte, leading to a catastrophic short circuit which results in combustion of the electrolyte [13, 16, 17]. Another approach, as mentioned in Chapter 1.2, involves exploiting AM to create 3D, architected electrodes, like those shown in Figure 1.3. In this fashion, the thickness of the the cathode and anode can be increased while maintaining a constant spacing between and into the electrodes, keeping lithium ion diffusion paths short and thus avoiding power losses [14, 48]. The complex electrode geometries enabled by AM processes, however, necessitate the development of a solid electrolyte that can be conformally coated onto and inbetween them [14, 48]. Polymer electrolytes like poly(ethylene oxide) (PEO), have been natural candidates for this role since the 1970's, when Fenton et al. discovered the ability of the Lewis-basic ether oxygens in the backbone to coordinate metal alkali ions [18], and Armand realized their potential application as electrolytes [19, 43]. PEO varieties like poly(ethylene glycol) diacrylate (PEGDA) can also be UV-cured in situ [20], which makes them amenable to use as conformal electrolytes in 3D batteries.

PEO and its analogues, however, present some issues. Research has demonstrated that the ionic conduction mechanism of alkali ions occurs through a combination of ion-hopping and the segmental motion of the PEO backbone in amorphous regions of the bulk polymer [49–51]; as segments of the PEO chain relax, they pass lithium ions from coordination shell to coordination shell. PEO-based electrolytes, however, suffer from low ionic conductivity, typically orders of magnitude below the 10^{-3} S/cm threshold sufficient for application in everyday batteries [52]. This is generally understood to be a function of the strong solvation of lithium ions by the PEO backbone; though this behavior is what allows PEO to dissolve lithium salts, it also presents a pronounced activation barrier to motion of lithium ions through the bulk polymer [16]. Mechanisms that seek to improve the segmental motion of chains in PEO-based electrolytes to improve ion transport generally come at the cost of reduced modulus, which is important for dendrite suppression [53]. Moreover, the anion remains loosely bound and more mobile than the lithium cation, which leads to a low cation transference number [16].

Much effort has been devoted to overcoming the tradeoff between ionic conductivity and modulus, in addition to the issue of low cationic transference number. Approaches have involved the use of lithium salts with low lattice energies and anions with diffuse charge distributions, like lithium bis(trifluoromethanesulfonylimide) (LiTFSI) [54]; the use of liquid and ceramic nanoparticle plasticizers to lower the glass transition temperature, T_g, to ensure the polymer remains as amorphous as possible [54–58]; attaching the anion to the polymer backbone to make single ion conductors [17, 59]; and the use of Lewis-acidic anion trapping molecules as additives or directly incorporated into the polymer backbone to slow the movement of anions relative to lithium cations [60–65].

Theoretical and computational work has suggested that a polymer with Lewis-acidic moieties on the backbone as opposed to the Lewis-basic ether oxygen moieties on the PEO backbone could improve both lithium-ion conductivity and transference number, and do so without depending on polymer segmental motion [66]. Such Lewis-acidic sites would preferentially coordinate the anion and leave the lithium cation relatively unhindered [66]. Though previous investigations have incorporated such moieties into polymer structures, they have not created polymer electrolytes with the intent to mediate ion solvation and conduction primarily through these Lewis-acidic groups [60–63].

We first attempt to create ionically conductive polymers with no ether-oxygen moieties in the backbone by hydroboration of polybutadiene with a common dialkylborane reagent, 9-borabicyclo(3.3.1)nonane, which results in a polymer we have named poly(9-BBN). Subsequent addition of n-butyllithium (n-BuLi, our "salt") provides mobile cationic charges and completes fabrication of the electrolyte. Such electrolytes have been demonstrated to be reliably ionically conductive, with certain exemplars approaching a normalized conductivity of 10^{-7} S/cm at 80°C. Moreover, these electrolytes are expected to exhibit near unity-transference given the covalent character of the interaction between the carbanions and the tricoordinate boron centers on the backbone, which was confirmed via solid state nuclear magnetic resonance (SSNMR). However, controlling the conductivity of the electrolytes as a function of salt concentration has proven to be an intractable problem: multiple samples from the same batch can demonstrate conductivities that differ from each other by orders of magnitude, and this problem persists even after controlling for various factors such as chemical quality and processing. This chapter describes the fabrication of these electrolytes and the challenges this presented and discusses what information could be meaningfully gleaned from the conductivity data obtained. Understanding what works and what does not in these systems provided the insight needed to design the more controllable electrolytes discussed in Chapter 4.

3.2 Fabrication of electrolyte films and measurement of their ionic conductivity

Unless otherwise stated, all chemicals were obtained from commercial suppliers and used as received. Solvents were dried by passage through activated alumina columns, degassed by sparging with Ar and stored over activated 3 Å molecular sieves under inert atmosphere, or anhydrous solvents were purchased directly from commercial suppliers. All glassware was oven dried for at least 12 h prior to use and synthetic manipulations were performed under a nitrogen or argon atmosphere inside a VAC/ATM glovebox (<1 ppm O_2). Deuterated solvents were purchased from Cambridge Isotope Laboratories, degassed by three freeze-pump-thaw cycles, and stored over activated 3 Å molecular sieves under inert atmosphere.

WeidentifiedatargetLewis-

acidicpolymersystembasedonthefacilehydroboration

ofpolybutadiene[67],

asshowninFigure3.1A.Wechosethecommerciallyavailable 9-borabicylo(3.3.1)nonane (9-BBN) as our borane. Poly(9-BBN) is synthesized by first recrystallizing commercial 9-BBN from 1,2 dimethoxyethane according to a purification procedure from the literature [68]. Polybudatiene (M_w = 1800, cis > 98%) from Sigma-Aldrich was degassed via three freeze-pump-thaw cycles and subsequently brought into a VAC/ATM glovebox with a nitrogen atmosphere. To a 20mLscintillationvialequippedwithamagneticstirbarwasaddedtherecry stallized 9-BBN dimer (225 mg, 0.92 mmol, 1 equiv) and commercial polybutadiene (100 mg, 1.85 mmol of olefin, 1 equiv). 10 mL of n-Hexane was added and the reaction mixture was stirred at 50°C. The 9-

BBN dissolves as the reaction proceeds to give a clear, colorless solution, which was stirred for at least 12 hours for the reaction to come to completion. After stirring for at least 12 hours, ^{11}B NMR indicated complete conversion of the 9-BBN dimer (δ 28 ppm) to the poly(9-BBN) (δ 88 ppm) (Figure 3.2).

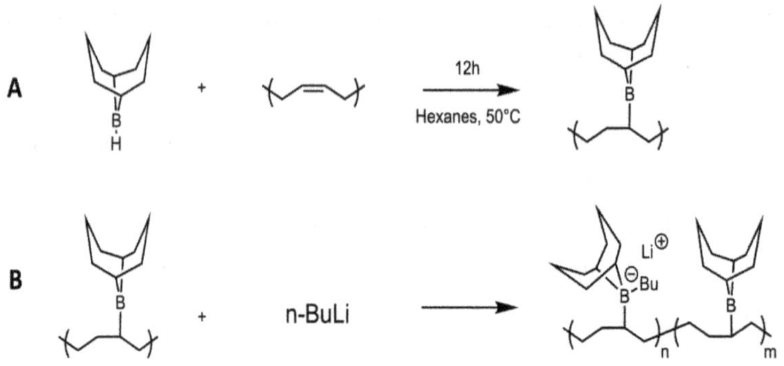

Figure 3.1: Reaction mechanisms for fabrication of polyborane electrolytes. (A) Synthesis of poly(9-BBN) via hydroboration of polybutadiene with 9-BBN. (B) Synthesis of final electrolyte via addition of n-BuLi to poly(9-BBN).

Trialkylboranes readily interact with carbanions via a Lewis acid/base interaction to form 4-coordinate tetraalkylborates (Figure 3.1B). A variety of small molecule examples are known[69], and the ^{11}B NMR shiftsof trialkylboranes and tetralkylborates are distinct and diagnostic, allowing for their easy identification. To the hexane solution of the of the poly-9BBN was added n-BuLi (solution in hexanes, 1.1 equiv relative to boron). Immediate precipitation of a white solid occurred, and the reaction mixture was stirred for 8 hours. The solid was isolated by vacuum filtration and washed with n-hexane and dried in vacuo. Solid state NMR confirmed both the formation of the 4-coordinate borate (^{11}B NMR δ -18 ppm) and the presence of Li cation (^{7}Li δ 1.4 ppm) (Figures 3.3 and 3.4).

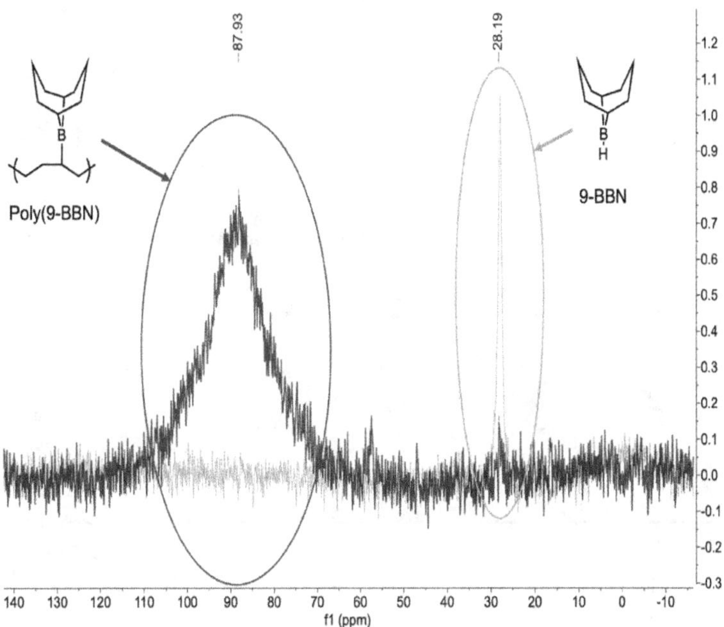

Figure 3.2: ^{11}B NMR spectra of 9-BBN dimer and poly(9-BBN). 9-BBN dimer exhibits a single peak at δ 28 ppm. Once consumed after hydroboration of polybutadiene, the only visible peak is that of tricoordinate boron, or trialkylborane, which sits at δ 88 ppm.

The solid obtained after reacting 1 equiv of n-BuLi with 1 equiv of poly(9-BBN) by boron could not be processed into a usable form for measurement of ionic conductivity by electrochemical impedance spectroscopy (EIS). Lower loadings, $r = mol_{Li}/mol_B$, and a multi-step fabrication process were required to allow for castingofthefinalelectrolyteproductintofilmsusableforEIS(Figure3.5).
Addition of n-BuLi directly to a solution of poly(9-BBN) in hexane, even at low loadings,

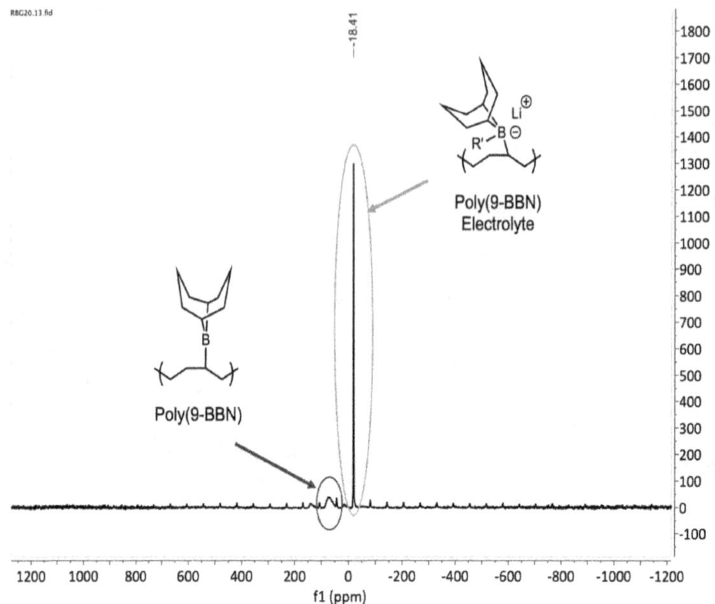

Figure 3.3: ^{11}B SSNMR spectrum of poly(9-BBN) polyelectrolyte formed after addition of n-BuLi to poly(9-BBN). Synthesis of final polyelectrolyte complex is confirmed by large upfield shift to the characteristic tetraalkylborate peak at δ18ppm. The shift also confirms the strength of the interaction between the butyl carbanion and the boron center, suggesting an immobile carbanion.

r = 0.125, 0.25, resulted in visible precipitation of the polyelectrolyte.

Though the polyelectrolyte remained suspended in hexane, it was not possible to ensure uniform and consistent drop casting of the electrolyte from this suspension. In order to produce electrolyte films for EIS, a 1-mL syringe was used to cast 0.15-0.2 mL of a solution of poly(9-BBN) in hexane with a 0.25 M concentration by boron onto stainless steel CR 2032 spacers. A 1-μL syringe was then used to add either 32 or 16 μL of 0.3 M n-BuLi in hexane directly onto the solution of poly(9-BBN) cast on the spacers to obtain loadings of r = 0.25 and r = 0.125, respectively. After allowing the cast films to dry for at least 24 hours in an N$_2$ glovebox atmosphere and then vacuum drying them in the glovebox antechamber for another 2 hours, polymer electrolyte films were obtained (Figure 3.5B). The films were then assembled into a custom apparatus fabricated from a PTFE block for impedance measurements, as demonstrated in Figure 3.5A. Bores the diameter of CR 2032 spacers were carved into the block, and steel rods obtained from McMaster were cut and machined to fit

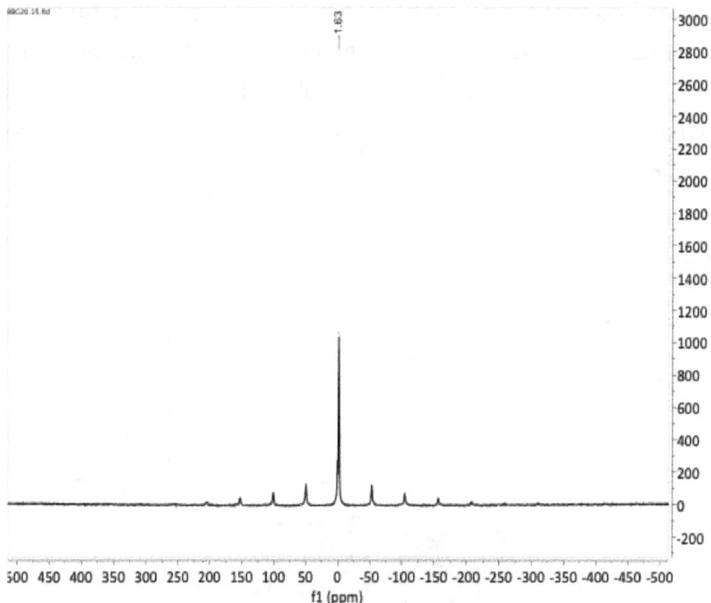

Figure 3.4: ^{7}Li SSNMR spectrum of poly(9-BBN) polyelectrolyte confirming the presence of lithium cations (δ 1.4 ppm) in the final electrolyte product.

snugly into the bores. Spacers with the cast and dried electrolyte films were loaded into the bores by first wrapping the end of a steel rod with Kapton tape (to ensure a tighter seal between the wall of the bore and the rod) and inserting it into one of the bores. Two-sided copper tape with a conductive adhesive was then placed on the end of the steel rod inside the bore, and the spacer with cast electrolyte was placed with its bare face on top of the tape. Another spacer was then placed on top of the electrolyte film, and the assembly was then sealed by inserting another steel rod wrapped in Kapton tape through the opposite end of the bore, with a piece of double-sided copper tape on its inserted end to ensure good electrical contact with the spacer. Two strips of copper tape were then affixed to the ends of the steel rods on each side, allowing for the connection of leads with alligator clips.

The entire apparatus was then removed from the N_2 glovebox and brought to a sealable oven with variable temperature control and a port allowing for external connection to a BioLogic potentiostat for EIS. The films were annealed in the oven at 80°C for 2h, and impedance measurements were subsequently taken from 70°C to 30°C in decreasing 10°C intervals, allowing the films 2h in between measurements

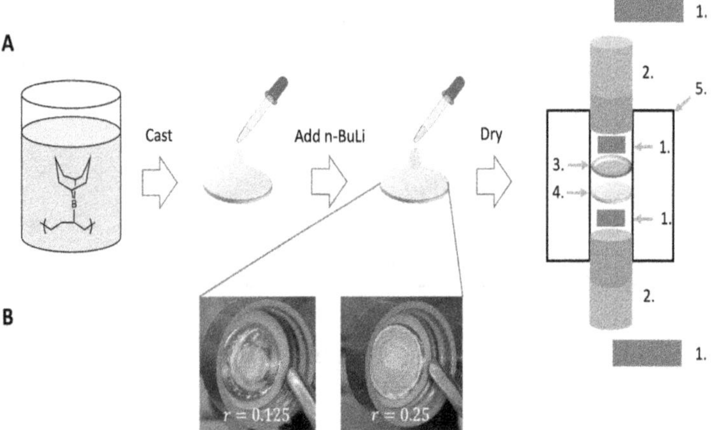

Figure 3.5: Schematic demonstrating fabrication process for poly(9-BBN)-based electrolyte films. (A) The fabrication process begins with a 0.25 M solution by boron of poly(9-BBN) in hexane. Using a 1-mL syringe, 0.15 mL is cast onto a CR 2032 spacer. Immediately after, a precise amount of 0.3 M solution of n-BuLi in hexane is added to the cast solution on the spacer using a 1-μL syringe. 32 μL is added to obtain a film with $r = 0.25$, or 16 μL to obtain a film with $r = 0.125$. The film is then allowed to dry for 24 hours in the N_2 glovebox atmosphere, after which it is dried another 2 hours under vacuum in the glovebox antechamber. The cast films are then assembled into a custom impedance apparatus made from a block of PTFE (5.). The spacers (4.) with the cast films are placed into bores drilled into the PTFE block, where another blank spacer (3.) is placed on top, and the spacer-electrolyte-spacer stack is sandwiched between two steel rods (2.) wrapped in Kapton tape and machined to fit snugly into the bores. Double-sided copper tape (1.) is affixed between the rods and spacers

to ensure good electrical contact, and at the end of the rods to allow connection to external leads.

to properly equilibrate at each temperature. An applied AC signal of 20 mV was employed for impedance measurements, which were carried out from 1 MHz to 500 mHz. Ionic resistances were extracted from the impedance spectra by performing simpleVoigtelementcircuitfitsonthesemicircularfeatures,andioniccond uctivities calculatedaccordingtotheformula $\sigma = L/RA$, where A istheareaofthefilms, L the thickness of the films, and R the bulk ionic resistances extracted from the impedance spectra. A is calculated from the area of the CR 2032 stainless steel spacers onto whichthe polymerelectrolytesare cast, and L isobtained bymeasuringthe thickness of the spacer-electrolyte-spacer stack after impedance measurement and subtracting from it the pre-measured thicknesses of the spacers in the stack.

3.3 Results

Polymer electrolyte films based on poly(9-BBN) were fabricated with r = 0.125 and r = 0.25 according to the procedure outlined in the previous section and Figure 3.5. ^{11}B SSNMR spectra confirm that addition of n-BuLi results in attachment of butyl carbanions to the boron centers in poly(9-BBN), which is readily identified by the pronounced upfield shift to δ-18 ppm, as seen in Figure 3.3. ^{7}Li SSNMR reveals the presence of lithium cations at δ 1.4 ppm, and this result, taken together with the [11]B spectrum, confirms the successful fabrication of the polymer electrolyte films.

Afterthefilmswerecast, dried, andassembledintothecustomimpedanceapparatus, EIS was performed on the electrolytes. The bulk ionic resistances, R were extracted from two sets of polymer electrolyte films, one with r = 0.125 and the other with r = 0.25, and these were in turn used to calculate the ionic conductivities of each film in each set. Figure 3.6 shows the ionic conductivities of the films in each set in a semilogarithmic Arrhenius plot against 1000/T. There are four films at r = 0.125, and six films at r

= 0.25. The highest ionic conductivity achieved is 5.7×10^{-8} at 70°C, in a film with $r = 0.125$. The variability of ionic conductivity within films of the same set is high, with the curves at each concentration spanning roughly two orders of magnitude.

To better understand the potential mechanism of ionic conduction at play, we performed simple fits of the Arrhenius equation:

$$\sigma = A exp \frac{Ea}{RT} \tag{3.1}$$

to the conductivity data of each film in each set. A is the pre-exponential factor, R is the universal gas constant, and Ea is the activation energy, here associated with the barrier to ionic motion in our polymer electrolyte system. The extracted Ea and A are listed Table 3.1.

3.4 Discussion

Chemical characterization of the films via SSNMR reveal the successful synthesis of the poly(9-BBN)-based electrolyte. The ^{7}Li SSNMR spectrum (Figure 3.4) confirms the presence of lithium cations in the final product, and the ^{11}B SSNMR spectrum (Figure 3.3) reveals a strong interaction between the butyl carbanion and the boron center, as indicated by the shift of the main isotropic peak upfield to δ -18 ppm. Given how tightly bound the carbanion and boron centers are, it could be

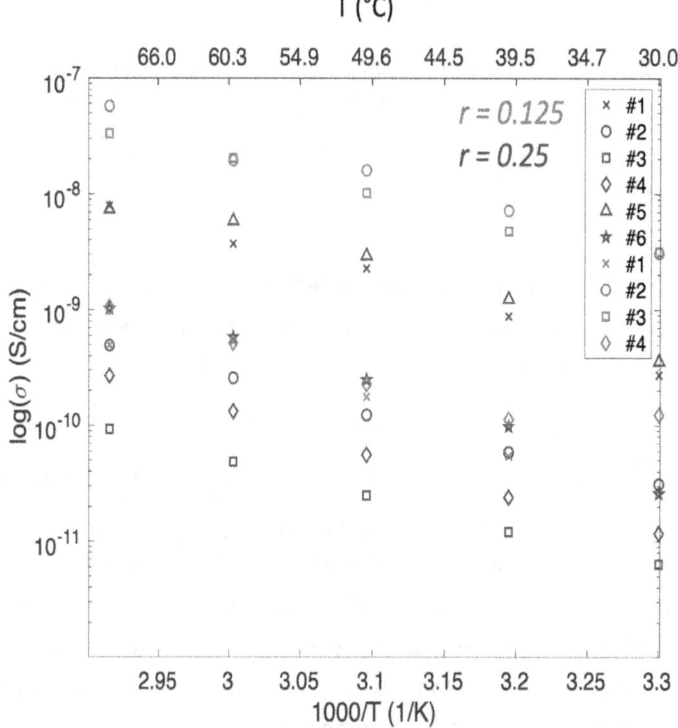

Figure 3.6: Ionic conductivity, σ, of poly(9-BBN)-based electrolyte films plotted against 1000/T on a semilogarithmic Arrhenius plot. There are two sets of data, one at r = 0.125, in red, with four examplars; and another at r = 0.25, in blue, with six exemplars. The conductivity of each exemplar was measured from 70°C to 30°C in decreasing 10°C intervals. There is large variability of ionic conductivity, even within the same set, which precludes any meaningful conclusion about the relationship between concentration and ionic conductivity.

$r =$ mol_{Li}/mol_B	Film No.	$E_d (kJ/mol)$	$A(S/cm)$
0.125	1	71.16	49.45
0.125	2	59.12	49.95
0.125	3	53.2	4.200
0.125	4	49.81	0.0333
0.25	1	71.02	560.6
0.25	2	60.41	0.7578
0.25	3	58.31	0.0685
0.25	4	68.75	7.7292

| 0.25 | 5 | 66.3 | 124.1 |
| 0.25 | 6 | 79.3 | 1458 |

Table 3.1: Activation energy, E_a, and pre-exponential factor, A, from least squares fit of simple Arrhenius equation to temperature-dependent conductivity of electrolyte films at all $r = mol_{Li}/mol_B$ in Figure 3.6.

surmised that the electrolyte we created has features akin to a single ion conductor [60], though more investigation would be needed to confirm this hypothesis.

When examining the ionic conductivity of the poly(9-BBN)-based electrolyte films at $r = 0.125$ and $r = 0.25$, it is apparent that the variability between films in the same set precludes any attempt to relate ionic conductivity to salt concentration. The difference between the best and worst performing cells in each set spans close to two orders of magnitude, and there is no discernible trend in ionic conductivity as concentration is varied. Moreover, the data presented in Figure 3.6 were taken after a number of adjustments were already made to the fabrication process in an attempt to control this spread. Potential causes for variability in the data, such as casting films directly from a heterogeneous suspension of poly(9-BBN) polyelectrolyte; waiting for poly(9-BBN) solution drop cast onto spacers to dry before adding n-BuLi; and quality of constituent chemicals like 9-BBN, polybutadiene, and n-BuLi were already addressed when the data in Figure 3.6 was collected. Another subsequent set of data was obtained from a new set of films, four each at $r = 0.125$ and $r = 0.25$, all according to the same refined procedure outlined in Figure 3.5. The ionic conductivities of these films also demonstrated the same wide swing over roughly two orders of magnitude, and no meaningful conclusions could be made about the relationship between ionic concentration and conductivity (Figure 3.7).

Having already accounted for chemical and processing issues that could have caused the large spread in measured ionic conductivity, the only aspect left to consider was the impedance apparatus itself. The choice to use a custom setup for EIS, as opposed

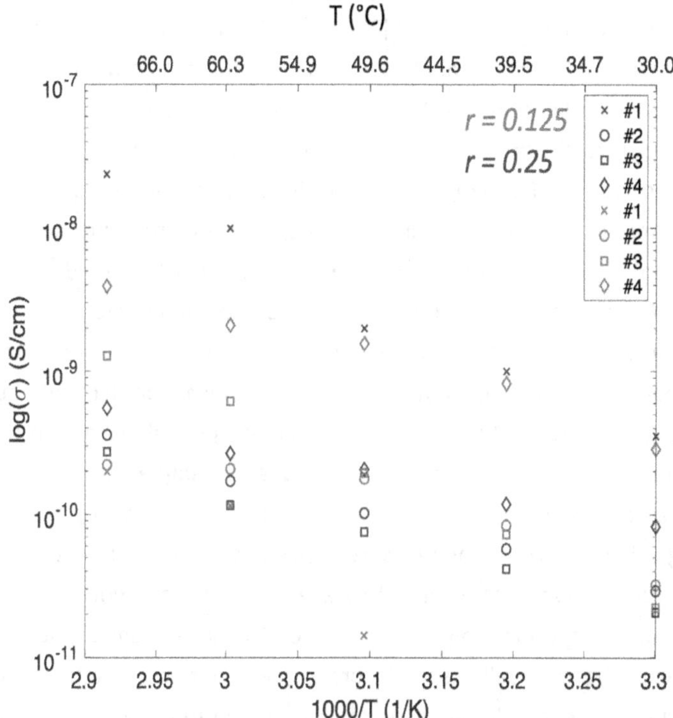

Figure 3.7: Ionic conductivity, σ, of a second set of poly(9-BBN)-based electrolyte films plotted against 1000/T on a semilogarithmic Arrhenius plot. There are eight electrolyte films, four with $r = 0.125$ (in red) and four with $r = 0.25$ (in blue). Like the data in Figure 3.6, there is large variability of ionic conductivity, even within the same set, precluding any meaningful conclusion about the relationship between concentration and ionic conductivity.

to a more typical coin cell arrangement like the one pictured in Figure 4.1 of the following chapter, followed from difficultiesin measuring ionic conductivity of films in coin cells. The pressure of sealing coin cells, in addition to temperature cycling during EIS, frequently resulted in cell shorting. To account for the electrolyte films being pushed out of place because of pressure and temperature, thin shims were punched out from PTFE sheets and placed on top and around the films to prevent shorting when assembled into coin cells. Though this solved the shorting issue, poor contact between the top spacer and the film was poor, as evidenced by the appearance of a small semicircle at very high frequencies (Figure 3.8).[70] This did not allow for proper assignation of circuit elements to features in the acquired impedance spectra as outlined in chapter 2.3. Moving to the custom apparatus allowed us to circumvent the shorting and contact issues by removing the need for a PTFE shim, but we surmise that the method of sealing the films in the apparatus involved applying different pressure to different cells within the same set, which could have affectedthesizeofthesemicircularfeaturesusedtoextractionicbulkresist ancesand resulted in the wide swings observed in ionic conductivities. Mechanical pressure is known to affect electrical contact and thus impedance spectra [70], so to control for this possible effect on our electrolyte films, we needed to ensure that each sample was subject to the same pressure when assembled for EIS. This meant a return to using coin cells, where the pressure applied by a crimper to seal them can be observed and controlled. However, modifications would have to be made to the electrolytes themselves to make them capable of surviving the coin cell assembly process without the resultant cells shorting. It was decided that the most direct route to more physically resilient cells would be to introduce a crosslinking molecule that could bind to unreacted olefins on the polybutadiene backbone to create a crosslinked polymer electrolyte network. This could make the films mechanically robust enough to withstand the pressure associated with coin cell assembly and the temperatures experienced during EIS experiments in the oven, while preserving as much of the novel qualities of the electrolyte as possible. This adjustment and the results are discussed in Chapter 4.

The only potentially meaningful piece of information that could be gleaned from the ionic conductivity data in Figure 3.6 was obtained from fits of the Arrhenius equation to the data obtained from each film, which is presented in Table 3.1. For all of the films at each concentration, E_a was found to be in the range of 50-70 kJ/mol, suggesting a potential similarity in the ionic conduction mechanism across all of the films despite the great inconsistency in measured ionic conductivity data.

3.5 Summary

In summary, we attempted to create a novel polymer electrolyte system with no ether oxygen atoms based on polybutadiene to drive ionic solvation and conduction through interaction with anions as opposed to cations. We created this polymer electrolyte by first hydroborating polybutadiene with a common borane, 9-BBN, to create a polymer with Lewis-acidic moeities on the backbone, poly(9-BBN). We then introduced ionic species to poly(9-BBN) by adding a precise amount of an organolithium reagent, n-BuLi. The synthesis of the resulting poly(9-BBN) electrolyte was confirmed via ^{7}Li and ^{11}B SSNMR, with the latter suggesting the

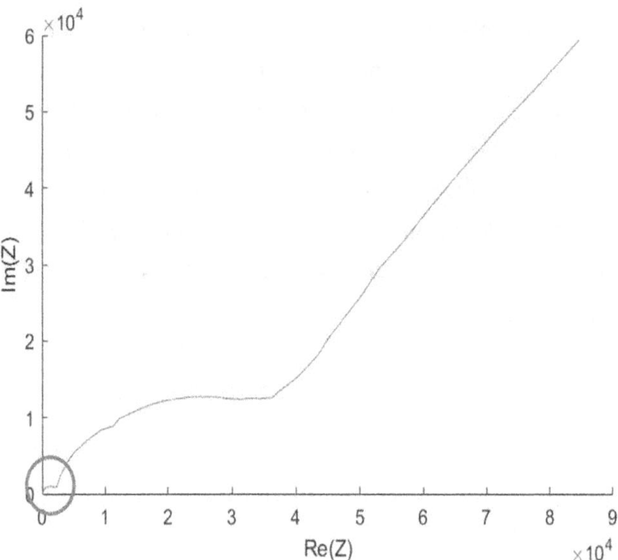

Figure3.8:
Impedancespectrumtakenfromanearlyelectrolytefilmincoincellwith poor electrical contact because of PTFE shim. Note the small semicircle circled in red at very high frequency. The appearance of such bumps have been attributed to poor contact between solid electrolytes and blocking electrodes like stainless steel spacers [70].

creation of a single-ion conducting polymer electrolyte. EIS was performed on two sets of electrolytes, one with $r = 0.125$ and another with $r = 0.25$, to measure their ionic conductivity, but this did not afford clarity on the relationship between ion concentration and ionic conductivity. The large variability of measured ionic conductivity between films at the same r did not provide any meaningful insight to potential ionic conduction mechanisms. Attempts to control for this inconsistency across films in the same set were made by addressing various aspects of the synthesis process, like the quality of the chemical ingredients used and the procedure for casting the electrolyte films. When the problem persisted, it was determined that the custom apparatus used to measure impedance—originally designed to circumvent issues with electrical contact and shorting from early EIS experiments with films in coincells—didnotallowforuniformapplicationofmechanicalpressuretoelectrolyte films when assembled. It was decided that a return to using coin cells would allow for tighter control over the mechanical pressure applied to the films, but adjustments would need to be made to the electrolytes to enable them to survive the coin cell assembly process. Introducing a UV-curable crosslinking molecule could make the films more mechanically resilient while preserving the physically novel features of the electrolyte. This approach and its results are covered in the following chapter.

C h a p t e r 4

EXPLORATION OF IONIC CONDUCTION IN POLYBORANE-BASED POLYMER ELECTROLYTES FOR LITHIUM AND LITHIUM-ION BATTERIES

This chapter has been adapted from:

F. J. Villafuerte et al. "Exploration of Ionic Conduction in Polyborane-based Polymer Electrolytes for Lithium and Lithium-ion Batteries (*In preparation*)".

4.1 Introduction

In this chapter we develop a variation of the novel polymer electrolyte system proposed in chapter 3, which similarly incorporates Lewis-acidic boron moieties onto a polybutadiene backbone in order to experimentally probe the possibility of mediating ionic solvation and conduction through these Lewis-acidic groups. This is achieved by hydroboration of polybutadiene using 9-BBN. The resulting poly(9BBN)-co-polybutadiene is treated with lithium tert-butoxide (LitBuO) as a salt, 1,4 butanediol diacrylate as a crosslinking molecule, and diphenyl(2,4,6trimethylbenzoyl)phosphine oxide (TPO) as a photoinitiator to produce a precursor resin, which is then drop cast onto PTFE spacers, UV-cured for 5 minutes, dried, and assembled into coin cells for electrochemical impedance spectroscopy (EIS) and into pans for differential scanning calorimetry (DSC) (Figure 4.1). The process of incorporating crosslinking molecules which cure the drop cast polymer electrolytes when exposed to UV light ensures that they are able to withstand the coin cell assembly and crimping process without shorting. We show that ionic conductivity (σ) of these polyborane based electrolytes (PBEs) as measured by EIS as a function of molar salt ratio, $r = mol_{Li}/mol_B$, does not track with their measured glass

transition temperatures, T_g or the activation energies, E_a, extracted from fitting the Vogel-Tammann-Fulcher (VTF) equation to the conductivity data. Morevoer, measurement of ionic conductivity of control PBE films without boron on the polybutadiene backbone confirms that the presence of Lewis-acidic boron groups is necessary for ionic solvation and conduction. Further analysis that compared the PBEs to a well-studied PEO-based electrolyte in the literature[1] through the calculation of a reduced conductivity, σ_r, to control for polymer viscosity and segmental motion, revealed that PBEs obtain optimal conductivity at higher salt concentrations than PEO, and that their ionic conductivities are far below that of PEO. We attribute these effects to the strong interaction between the Lewisacidic boron centers and the strongly Lewis-basic tert-butoxide anions, which limits ionic conductivity by suppressing motion of the anions, and by presenting a large activation barrier for motion of Li+, which is optimized at concentrations where the distance between the boron-anion centers is sufficiently small to increase the probability of a hopping event from one center to another.

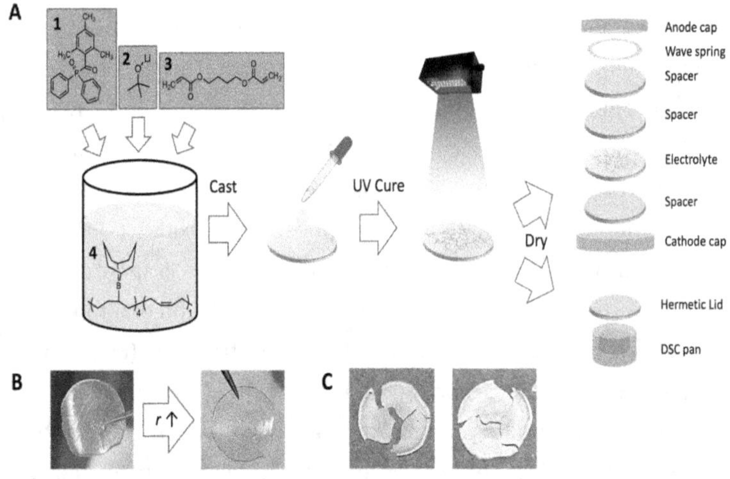

Figure 4.1: Synthesis of polyborane electrolyte films. (A) Demonstration of film synthesis. ProcessbeginswithcreationofPoly(9BBN)-co-polybutadiene(4), which is achieved through mixture of 9-BBN in THF with polybutadiene in THF at 50°C for at least 6h. TPO Photoinitiator (1) and 1,4 butanediol diacrylate crosslinker (3) are subsequently added, in addition to lithium tert-butoxide (2) at various concentrations. The mixture is then cast onto circular PTFE spacers and cured under UV light for 5 minutes. UV cured films are allowed to dry in glovebox atmosphere for a day before being dried in vacuum for at least 2 hours,

and are then assembled into coin cells for EIS measurements. This process is repeated directly in a DSC pan, which is then hermetically sealed for DSC experiments. (B) As the fraction of ionic species, $r = mol_{Li}/mol_B$, increases, the color of the PBE films becomes progressively more orange. All the films allow the passage of light. (C) Image of a control film fabricated with the same procedure, but with pure polybutatdiene as a base, as opposed to poly(9-BBN)-co-polybutadiene. Note that the film is differently colored on both sides and opaque, suggesting that ionic species were not thoroughly dissolved in the polymer matrix.

4.2 Experimental methods

Unless otherwise stated, all chemicals were obtained from commercial suppliers and used as received. Solvents were dried by passage through activated alumina columns, degassed by sparging with Ar and stored over activated 3 Å molecular sieves under inert atmosphere, or anhydrous solvents were purchased directly from commercial suppliers. All glassware was oven dried for at least 12 h prior to use and synthetic manipulations were performed under a nitrogen or argon atmosphere inside a VAC/ATM glovebox (<1 ppm O_2). Deuterated solvents were purchased from Cambridge Isotope Laboratories, degassed by three freeze-pump-thaw cycles, and stored over activated 3 Å molecular sieves under inert atmosphere.

Statisticalpoly(9-BBN)-co-polybutadienepolymersweresynthesizedbyfirstrecrystallizing commercial 9-BBN from 1,2-dimethoxyethane according to a purification procedure from the literature.[68] Polybudatiene (Mw = 1800, cis > 98%) from Sigma-Aldrich was degassed via three freeze-pump-thaw cycles and subsequently brought into a VAC/ATM glovebox with a nitrogen atmosphere. Approximately 100 mg of polybutadiene were pipetted into a 4 mL scintillation vial, and to the vial was added a precise volume of 0.5 M 9-BBN in THF stock solution to hydroborate 80% of the polybutadiene backbone. A magnetic stir bar was added to the vial, which was then sealed and placed on a magnetic hot plate set to 50°C and allowed to stir overnight. This process was repeated to make 8 vials of poly(9BBN)-co-polybutadiene. Solution state NMR was performed on samples of the poly(9-BBN)-co-polybutadiene using Varian 400, 500, and 600 MHz spectrometers or a Bruker Avance 400

MHz spectrometer to ascertain completion of the hydroboration reaction. ^{11}B NMR spectra were recorded with quartz NMR tubes (Wilmad 535-PP-7QTZ); more qualitative ^{11}B NMR spectra were recorded using borosilicate tubes (Wilmad WG-1000-8) and processed using the FID shift tool in MestReNova to remove the broad background signal from the glass. The final product was confirmed by observing the disappearance of the peak associated with the 9-BBN dimer (δ 28 ppm)[71] and the appearance of the trialkylborane peak (δ 85 ppm)[72] associated with the 9-BBN attached to the polybutadiene backbone.

After stirring, the vials were removed from the hot plate and allowed to cool to ambient temperature. To each vial was added a precise volume of a 0.5 M solution

of lithium tert-butoxide in THF to obtain specific molar ratios, $r = mol_{Li}/mol_B$, of 0.25, 0.33, 0.4, 0.5, 0.6, 0.75, 0.8, and 1. Halfofthevolumeofeachsolutionwasthen removedinvacuo, and toeachvialwasthenadded50 μL of1,4butanedioldiacrylate (crosslinker) and 0.2 mL of a solution of diphenyl(2,4,6-trimethylbenzoyl)phosphine oxide (TPO, photoinitiator) in THF with a concentration by mass of 2.5 mg per 0.1 mL of THF (See Figure 4.1A).

The solutions were then drop cast onto PTFE spacers 15.5 mm in diameter. The drop cast films were then cured under UV light for 5 minutes, and allowed to dry overnight in the glovebox atmosphere. All films were then dried under vacuum for at least 2 h in the glovebox antechamber to remove any residual solvent, removed from the PTFE spacers, sealed in glass vials, and moved to an Ar-filled glovebox, where the films were sandwiched between stainless steel CR2032 spacers and assembled into CR2032 coin cells for electrochemical impedance spectroscopy (EIS) as shown in Figure 4.1A. The stock electrolyte solutions were also used to drop cast films into hermetically sealable pans for DSC measurements. The process for curing and drying the films cast into the DSC pans was the same as that of those cast onto the PTFE spacers. These films were then sealed into a glass vial and moved to an Ar-filled glovebox, where they were hermetically sealed using a crimper.

The coin cells were then removed from the Ar-filled glovebox and brought to a sealable oven with variable temperature control and a port allowing connection to a BioLogic potentiostat for impedance measurements. The cells were annealed in the oven at 80°C for 2 h, and impedance measurements were subsequently taken from 80°C to 30°C in decreasing 10°C intervals, allowing the cells 2h in between measurements to properly equilibrate at each temperature. An applied AC signal of 80 mV was employed for impedance measurements, which were carried out from 1 MHz to 500 mHz. Ionic resistances were extracted from the impedance spectra by performing simple Voigt element circuit fits on the semicircular features, and ionic conductivities calculated according to the formula $\sigma = L/RA$, where A is the area of the films, L the thickness of the films, and R the ionic resistances extracted from the impedance spectra.[44, 45] A is calculated from the area of the PTFE spacers onto which the polymer electrolytes are cast, and L is obtained by measuring the thickness of the spacer-electrolyte-spacer stack after impedance measurement and subtracting from it the premeasured thicknesses of these same spacers in the stack. A set of three control films without boron for EIS measurements was also fabricated using the same procedure for the PBEs, exchanging the poly(9-BBN)co-polybutadiene for regular polybutadiene. The films were fabricated assuming poly(9-BBN)-co-polybutadiene as a base, and adding enough lithium tert-butoxide to the specifications of $r = 0.163$, 0.25, 0.5 PBE films.

Differential scanning calorimetry (DSC) experiments were carried out on a Discovery 25 DSC instrument from Thermal Analysis (TA) Instruments. Samples were taken through heat-cool-heat cycles covering a range from -40°C to 100°C at a ramp rate of 10°C/min and were allowed to equilibrate at each extreme. The initial heat cycle and equilibration served to precondition and anneal the polymer, and the DSC profiles and their first derivatives reported here and used for obtaining T_g were taken from the second heat cycle (Figure 4.5). A custom python program was developed to obtain T_g by numerically calculating the first derivative of the heating profiles and searching for minima.

An additional set of PBE films with $r = mol_{Li}/mol_B$ ratios of 0.25, 0.33, 0.5, 0.6, 0.75, 0.8, and 1 were fabricated for characterization via Solid State NMR (SSNMR) using cross polarization magic angle spinning (CPMAS). The films were packed and sealed into 2mm MAS rotors inside the same Ar glovebox used to prepare samples for EIS and DSC measurements. The rotors were then removed from the glovebox for SSNMR characterization. [11]B, ^{13}C, and ^{7}Li Solid SSNMR spectra were recorded on a Bruker Avance 500 MHz spectrometer using a Bruker CPMAS probe, where the rotors were spun at frequencies of 8-12 kHz to produce spectra with sufficient resolution for characterization. All NMR spectra were processed using MestReNova v12.0.

4.3 Results

Material Characterization of PBEs. SSNMR was employed to characterize the PBE films made according to the procedure outlined in the Experimental Methods and Figure 4.1. Films with $r = mol_{Li}/mol_B$ ratios of 0.25, 0.33, 0.5, 0.6, 0.75, 0.8, and 1 were fabricated and [11]B (Figure 4.2), [13]C (Figure 4.3), and [7]Li (Figure 4.4) spectra were taken from each. The [11]B spectra are arrayed in a stack plot from low to high values of r, and each spectrum exhibits the characteristic trialkylborane signature at δ 73 ppm,[72] which confirms the integration of the poly(9-BBN)-copolybutadiene into the crosslinked PBE films. SSNMR spectra exhibit artifacts known as spinning sidebands that are separated from the main, isotropic peak by integer multiples of the angular frequency at which the sample is being spun inside the MAS probe. These features are manifested in the [11]B spectra at δ 134 and δ 20 ppm, which correspond to a chemical shift of 9 kHz, the frequency at which the sample rotors were spun inside the CPMAS probe. Additional features separate from the sidebands arise at δ -4ppm, each with a corresponding shoulder that varies in intensity from sample to sample. The [7]Li spectra in Figure 4.4 are also arrayed in stack plot from lowest to highest r, with a large isotropic peak at δ 0 ppm, which is typical of Li cations.[73–75] The peaks on either side of the isotropic peak at δ

62, -62 ppm are spinning sidebands, separated from it by chemical shifts of 12 kHz, the speed at which the sample rotor was spun. The ^{13}C NMR spectra in Figure 4.3 are arranged in the same stack plot format as the ^{11}B and ^{7}Li spectra, with prominent peaks at δ 34, 31, 26, and 24 ppm, in addition to a smaller feature of interest at δ 68 ppm. The peaks at δ 34, 24 ppm are attributable to two different C-atom environments in the 9-BBN ring,[71] and the peaks at δ 31, 26 ppm to aliphatic C-atoms in the polybutadiene and 1,4 butanediol diacrylate backbones.[76–78] The peak at δ 68 ppm is attributable to residual THF that remains in the PBEs after the drying process.[79] There is one additional peak at δ 36 ppm that arises in spectra for $r \geq 0.75$ and becomes more prominent with increasing molar salt ratio, which may be due to C atoms in the tert-butoxide anions.

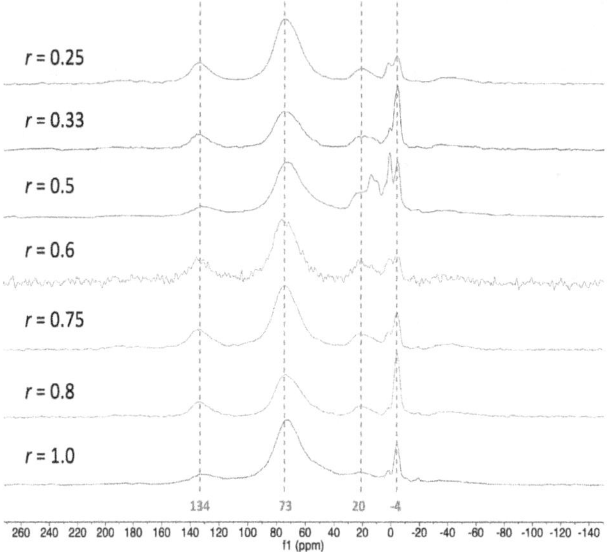

Figure 4.2: Stack plot of ^{11}B SSNMR spectra ascending in order of increasing molar salt ratio, r.

Thermal Characterization of PBEs. DSC was carried out to probe the thermal properties of the PBEs. Films with r of 0, 0.25, 0.33, 0.4, 0.5, 0.6, 0.75, 0.8, and 1 were fabricated according to the procedure outlined in the Experimental Methods and Figure 4.1 and subjected to the thermal cycling procedure also outlined in the Experimental Methods. T_g for each film was extracted from the second heating

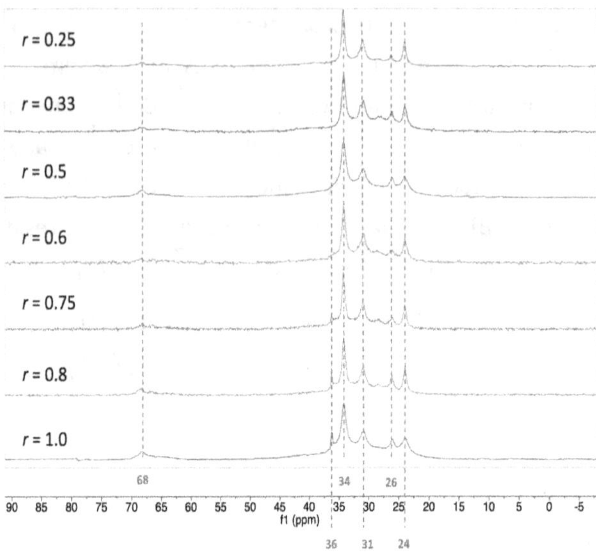

Figure 4.3: Stack plot of ^{13}C SSNMR spectra ascending in order of increasing molar salt ratio, r.

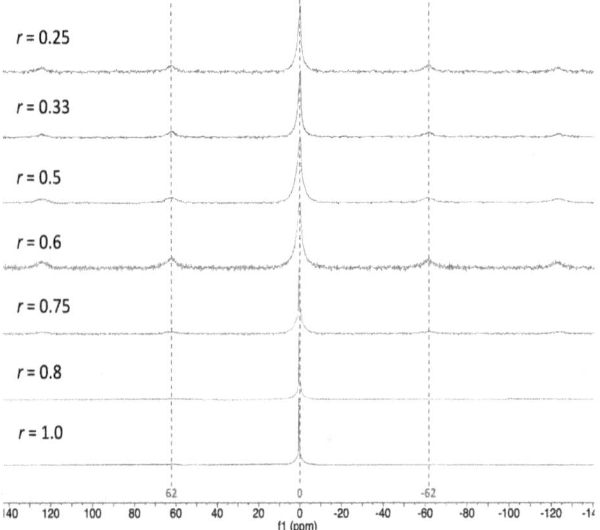

Figure 4.4: Stack plot of ^{7}Li SSNMR spectra ascending in order of increasing molar salt ratio, r.

cycle of each DSC run by finding minima of the first derivative of the heating profile in the region about the observed inflection point characteristic of a glass transition. (See Figure 4.5).

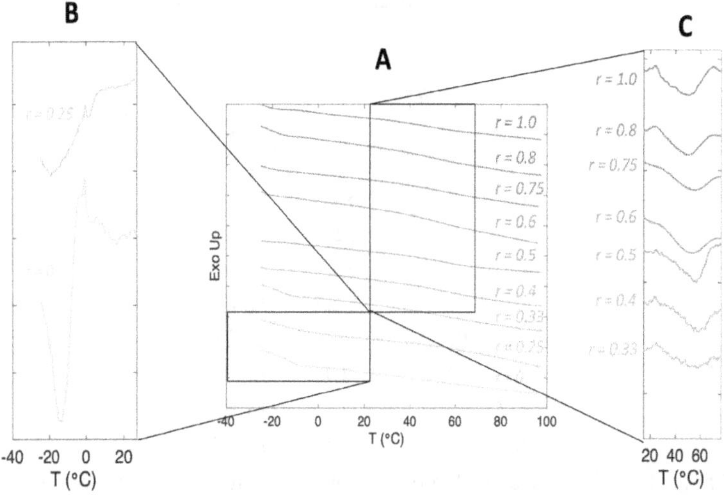

Figure 4.5: DSC heating curves and their first derivatives. (A) shows the DSC heating scans for exemplars at each r. The first derivatives were calculated to find minima indicative of a glass transition. (B) shows the first derivatives taken for films at $r = 0$, 0.25, which show the minima indicative of T_g just above -20°C. Plot C shows the first derivatives taken for films at $r \geq 0.33$, which showed minima indicative of T_g at higher temperatures, in a range from roughly 50 - 60°C.

The extracted T_g values collected in Table 4.1 are averages from sets with $n \geq 3$, and are plotted along with their standard deviations in Figure 4.9A. As r increases from 0 to 0.33, T_g jumps from from -13.3°C to 53.8°C, and thereafter plateaus at values between 49°C and 60°C. The set of films at $r = 0.33$ exhibited two T_g signatures, one at -19.35°C, and another at 53.8°C, both of which are reported in Table 4.1. The higher value was chosen for the analysis performed in this work. The implications of this are handled in the discussion.

Electrochemical Characterization of PBEs. Polymer electrolyte films with varying $r = mol_{Li}/mol_B$ ratios of 0.25, 0.33, 0.4, 0.5, 0.6, 0.75, 0.8, and 1 were fabricated from solutions of poly(9-BBN)-co-polybutadiene in THF with lithium tertbutoxide, 1,4 butanediol diacrylate, and diphenyl(2,4,6-trimethylbenzoyl)phosphine oxide (TPO) according to the procedure described in the Experimental Methods and outlined in Figure 4.1A. These films were then assembled into coin cells for

$r =$ mol_{Li}/mol_B	$T_g(^\circ C)$	Std. Dev.
1.0	49.5	±0.21
0.8	51.0	±1.97
0.75	55.0	±0.97
0.6	51.2	±1.64
0.5	55.0	±2.67
0.4	59.3	±2.04
0.33, #2	53.8	±5.28
0.33, #1	-19.4	±2.94
0.25	-16.8	±1.67
0	-13.3	±0.39

Table 4.1: T_g measured from DSC experiments as a function of $r = mol_{Li}/mol_B$, with standard deviation included

impedance measurements to obtain their ionic conductivity, σ, as a function of $1000/T$ and r. The results of these experiments are shown in Figure 4.6. Multiple PBE films for ionic conductivity measurements were made at each r, with $n \geq 3$ for each set, except for $r = 0.8$, where $n = 2$, and the ionic conductivities plotted in Figure 4.7 represent averages at each r. A plot of the experimentally obtained conductivity values as in Figure 4.6A with the standard deviation included for each point (except for $r = 0.8$, where the range plotted is the span between the two data points) can be found in Figure 4.7. Standard deviation is well within an order of magnitude for each r except $r = 0.4$, which we attribute to the difficulty of measuring conductivity at low ionic conductivities. More sample averaging would narrow this distribution, as is the case with $r = 0.25$ and $r = 0.33$, with $n = 6$, whereas $n = 3$ for $r = 0.4$. Given the trend of increasing conductivity as r increases towards $r = 0.75$, the average conductivities for $r = 0.4$ are reasonable.

Figure 4.6A is a semilogarithmic plot of measured ionic conductivity, σ, as a function of $1000/T$, with each color corresponding to a specific r.

At low salt fractions ($r < 0.5$), the increase in σ with increasing temperature is more pronounced, spanning close to an order of magnitude for $r = 0.25$. For $r \geq 0.5$, the conductivities do not exhibit the same marked increase with temperature, and the individual curves are more tightly clustered. Ionic conductivity achieves a maximum of 2.17 × 10^{-8} S/cm at a temperature of 80°C and $r = 0.75$. The trend in conductivity as a function of salt fraction is mirrored in Figure 4.6B, a semilogarithmic plot of σ against r.

Ionic conductivity increases close to two orders of magnitude as r increases from $r = 0.25$ to $r = 0.75$, after which it declines gradually. This trend is preserved

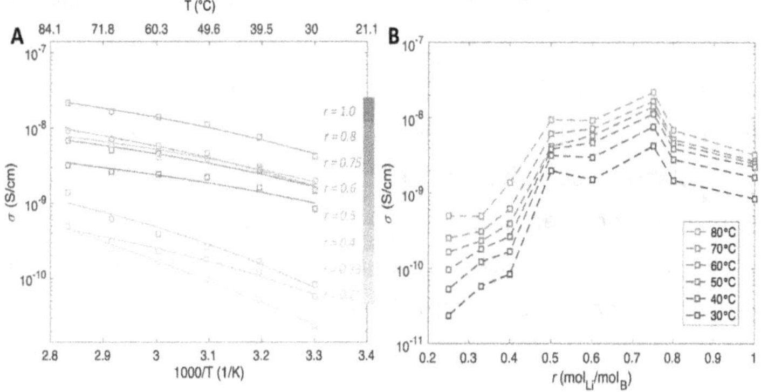

Figure 4.6: Electrochemical characterization of polymer electrolyte films at varying

$r = mol_{Li}/mol_B$. (A) Semilogarithmic plot of measured conductivity, σ, against $1000/T$ for each r. The conductivity increases with temperature, but markedly more at concentrations below $r = 0.5$. The lines correspond to fits of the VTF equation to each set of r. (B) Semilogarithmic Arrhenius plot of σ against r, with each curve corresponding to a specific temperature. Conductivity increases nearly two orders of magnitude from $r = 0.25$ to $r = 0.75$ across all temperatures.

across all temperatures.

EIS experiments were also carried out on the three control films targeted to molar salt ratios of $r = 0.163$, 0.25, 0.5 using the same parameters employed for the PBE films. EIS spectra from these films did not form the typical semicircular features on a complex plane plot, nor did they exhibit a plateau towards a finite value on a Bode plot, both of which are associated with a finite and measurable ionic resistance, which is needed for calculation of a finite ionic conductivity (Figure 4.8).

4.4 Discussion

The ^{11}B SSNMR spectra of the PBEs containing LiOtBu (Figure 4.2) reveal the presence of a resonance at δ -4 ppm, which could indicate interaction between the boron centers and the tert-butoxide anions, and thus a solvation mechanism driven primarily by the boron moieties on the polybutadiene backbone. The interaction of Lewis bases with the empty p-orbital in tricoordinate boron molecules typically producesachemicalshiftbelow0ppm.[72]Giventhatthereisnodiscernibl etrendin the qualitative intensity of this feature as r increases, it would be difficult to attribute this solely to a concentration-dependent anion-boron interaction and to discard other potential explanations. However, the lack of measurable ionic conductivity in the control films (Figure 4.8) demonstrates that: 1) Ether oxygen atoms in the

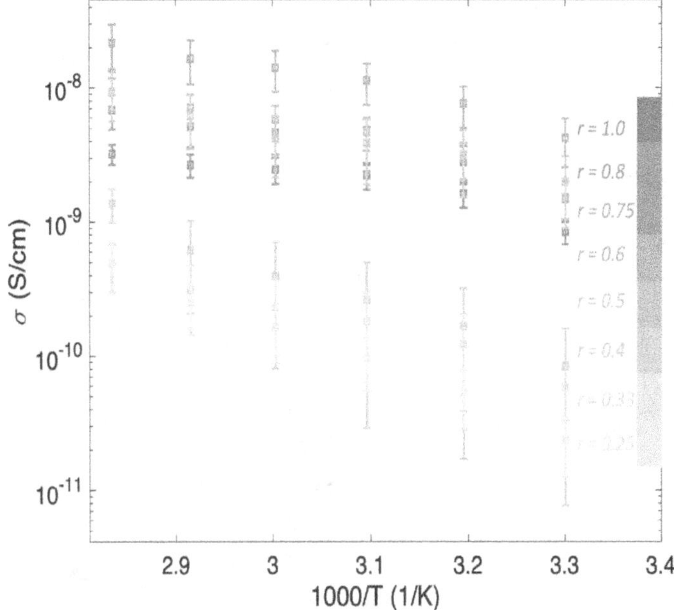

Figure 4.7: Plot of experimental conductivity at each r, as a function of $1000/T$, with error bars. Each data point represents a sample average, and the error bars represent one standard deviation above and below each average value except the points associated with $r = 0.8$, where the average comes from two samples, and the error bars represent the difference between the conductivities of each sample used for the average.

cross-linking 1,4 butanediol diacrylate molecules are not the primary driver of ionic solvation and conduction in the PBEs, and 2) Boron in the poly(9BBN)-copolybutadiene is necessary for ionic solvation and conduction. Visual inspection of the PBE and control films also lends more credence to the necessity of the boron groups for ionic conductivity: the PBE films in Figure 4.1B are transparent whereas the control films in Figure 4.1C are opaque, a qualitative indication that ionic species in the control films are not thoroughly solvated, a precondition for an ionically conductive electrolyte.

Ionicconductivityinpolymerelectrolytescanbeinfluencedbydifferentfactors, such as the segmental motions of the chains, ion mobility, and the concentration of ions, as discussed in chapter 2. The Vogel-Tammann-Fulcher equation is an empirical formula that relates polymer viscosity to ionic conductivity, by accounting for an

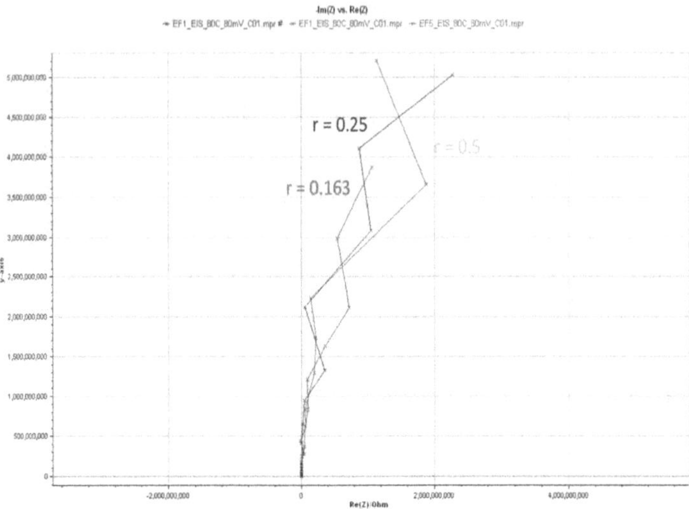

Figure 4.8: Plot of impedance spectra for control films with no boron moieties on the polymer backbone targeted to molar salt ratios of r = 0.16, 0.25, 0.5.

$r = mol_{Li}/mol_B$	$c\,(mol_{Li}/kg)$	$A\,(S/cm)$	$E_a\,(kJ/mol)$	R^2
1.0	3.90	$6.99 \times 10^-$	2.778	0.920
0.8	3.21	1.86×10^{-6}	3.076	0.965
0.75	2.82	6.76×10^{-6}	3.141	0.987
0.6	2.38	4.50×10^{-6}	3.713	0.982
0.5	1.76	1.75×10^{-7}	2.787	0.933
0.4	1.58	1.39×10^{-6}	4.668	0.962
0.33	1.32	2.90×10^{-7}	4.009	0.984
0.25	0.95	1.02×10^{-4}	16.16	0.995

Table 4.2: Activation energy, E_a, pre-exponential factor, A, and R^2 from least squares fit of VTF equation to temperature-dependent conductivity of electrolyte films at all r = mol_{Li}/mol_B. Li cation molar fraction per mass of polymer, c = mol_{Li}/kg, is also shown, which will be used to allow for comparison to PEO-LiTFSI systems.

SPE's glass transition temperature, T_g, below which an SPE becomes infinitely viscous, segmental motion ceases, and ionic conductivity is driven primarily by ion mobility and ion concentration:[38, 43, 80]

$$\sigma(T) = AT^{(-1/2)}exp\frac{-E_a}{R(T-T_0)}$$

(4.1)

where A is the preexponential factor, E_a is an activation energy, R is the universal gas constant, and T_0 is a reference temperature typically taken at least 50K below the empirically measured T_g,[38, 80] which for many polymers is understood as an idealized glass transition temperature known as the Vogel temperature, T_∞, at which the excess configurational entropy of the polymer or its free volume goes to zero.[38] The VTF equation was fit to the conductivity data at each r through a non-linear least squares method, where A and E_a are floating parameters extracted from the fits, and

T_0 is chosen to be $T_0 = T_g - 110K$, which produces good fits ($R^2 \geq 0.920$ for each

r) and allows for a later comparison to more well-studied PEO electrolyte systems. These fits are plotted along with the measured ionic conductivities in Figure 4.6A, and the extracted values for E_a and A, in addition to R^2 are listed in Table 4.2. The non-Arrhenius-like behavior of the conductivity as a function of temperature, in addition to how well the VTF equation fits the data, suggests that understanding the conductivity of the PBEs through the lens of their T_g is appropriate. In employing the VTF equation to understand ionic conductivity in PBEs, care must be taken to parse out trends in σ, T_g, and E_a as a function of r. Figure 4.9A is a plot of T_g as a function of r. From $r = 0.25$ to $r = 0.33$, there is a marked increase in T_g, from $-16.8 - 53.8°C$, and then a general plateau in T_g values thereafter in the range of $49.5 - 55.0°C$. It is generally understood that T_g in PEO-like systems increases monotonically with salt concentration per repeat unit of polymer, $r = mol_{Li}/mol_{EO}$, which is attributed to less flexible polymer chains due to weak crosslinking arising from the greater interaction of ether oxygen atoms on the polymer backbone with Li cations.[1, 38, 51, 80] In the

PBE systems, it is important to note that in the absence of ionic species when $r = 0$, the PBEs exhibit a T_g of $-13.3°$C, similar to the T_g of $-16.8°$C measured at $r = 0.25$. This suggests that the T_g of the PBEs at these low molar salt ratios is primarily a function of the crosslinked poly(9-BBN)co-polybutadiene network that forms the base of the PBEs. A threshold appears to be crossed at $r = 0.33$, at which point the T_g is driven higher by a sufficient concentration of ionic species. However, T_g remains relatively insensitive to further increases beyond $r = 0.33$, suggesting that the PBE transitions abruptly into a glassy, saturated salt regime.

Beyond r = 0.33, the insensitivity of E_a in Figure 4.9B to increases in molar salt ratio reinforces the idea of a transition to a glassy, salt-saturated regime, where changes in the amount of salt do not appreciably affect the ionic conduction mechanism at play. As r increases from $r = 0.25$ to $r = 0.33$, E_a decreases rapidly from 16.16 kJ/mol

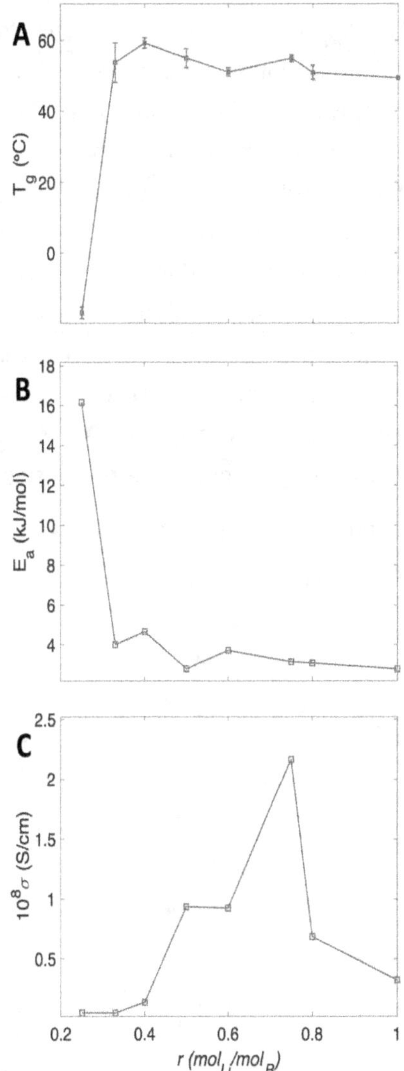

Figure 4.9: Trends in T_g, E_a, and σ as a function of r for PBE systems. (A) Plot of

T_g as a function of r. (B) E_a extracted from VTF fits of σ as a function of

r. Note that values of T_g and E_a stabilize for $r \geq 0.5$. (C) Plot of σ at $80°C$ (scaled by a factor of 10^8) against r. Unlike E_a and T_g, σ does not stabilize for $r \geq 0.5$.

to 4.009 kJ/mol, and then declines slightly before plateauing at values between 2.7 and 3.8 kJ/mol. This decrease in values for E_a suggests a decreased barrier to ionic motion, despite the corresponding increase in T_g, which is typically understood to imply an increased barrier to ionic motion in polymer systems like those based on PEO.[38, 51] It should be noted that the extracted values for E_a do depend in part on the choice of reference parameter, T_0, chosen for the VTF fits, so a thorough analysis requires considering the sensitivity of E_a to the choice of T_0. A sensitivity analysis of E_a from the VTF fits was carried out by fitting the experimental data with varying values of T_0. The results of this analysis are summarized in Figure 4.10. Though there are slight changes in the magnitude of the values for E_a from the fits, the general trend in E_a as a function of r is preserved, suggesting that the insensitivity of the ionic conduction mechanism to $r \geq 0.33$ is indeed a feature of the PBE system.

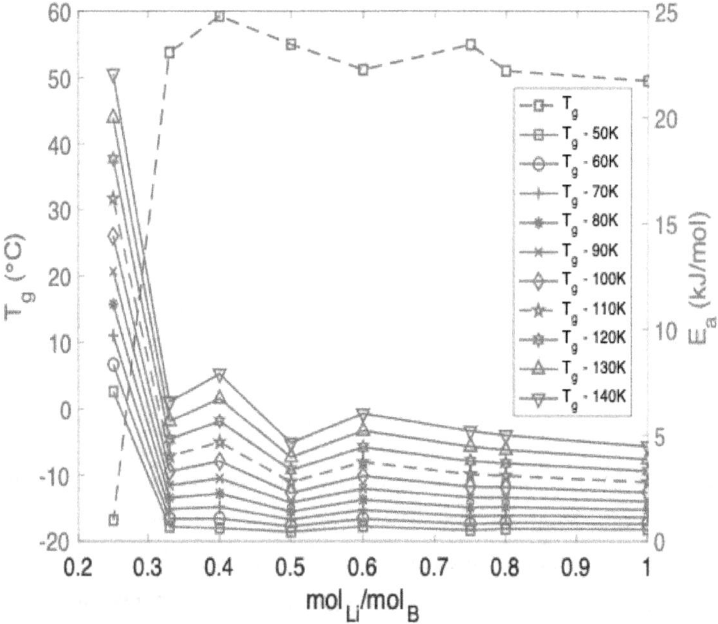

Figure 4.10: Analysis of sensitivity of E_a values extracted from VTF fits to changes in reference temperature, T_0. The VTF fits were performed with $T_0 = T_g - 50K$ to $T_0 = T_g - 140K$, in decreasing increments of $10K$.

As the T_0 varies, the general trend in E_a as a function of $r = mol_{Li}/mol_B$ is the same, decreasing and then plateauing for values of $r \geq 0.5$. The data set with dashed lines corresponds to $T_0 = T_g - 110K$, which was chosen for analysis and comparison to PEO.

This trend also persists when accounting for the fact that two T_g values were measured at $r = 0.33$, as noted in the results section. Choosing the lower of these two values,
$T_g = -19.4\ °C$, for the VTF analysis does not alter the trends observed in Figure

4.10, other than shifting the abrupt step increase in T_g values to a slightly higher

$r = 0.4$, and introducing an intermediate value for E_a of 11.36 kJ/mol at $r = 0.33$ between 16.16 kJ/mol at $r = 0.25$ and 4.668 kJ/mol at $r = 0.4$ (See Figure 4.11).

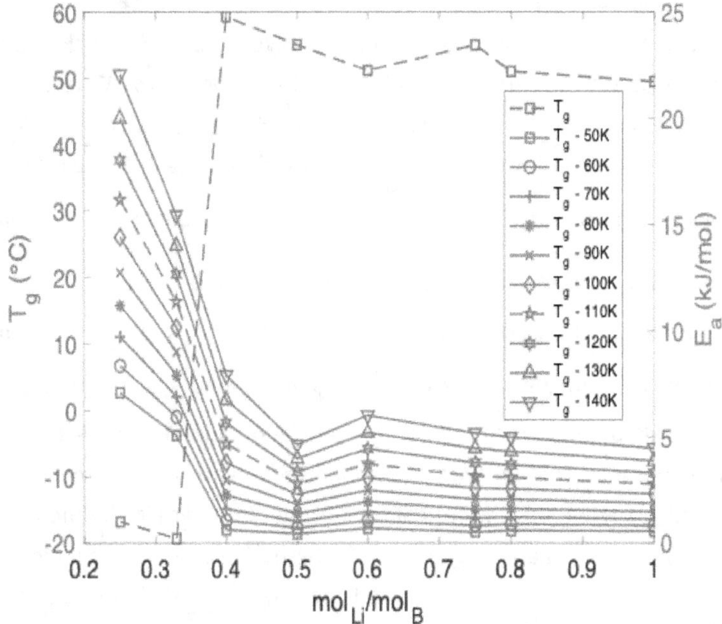

Figure 4.11: Analysis of sensitivity of E_a values extracted from VTF fits to changes in reference temperature, T_0, with $T_g = -19.4°C$ for $r = 0.33$. The VTF fits were performed with $T_0 = T_g - 50K$ to $T_0 = T_g - 140K$, in decreasing increments of $10K$, and where the T_g chosen at $r = 0.33$ was

the lower of two values measured, T_g = -19.4°C or 253.6K. As the T_0 varies, the general trend in E_a as a function of

$r = mol_{Li}/mol_B$ is the same, decreasing and then plateauing for values of $r \geq 0.5$. The data set with dashed lines corresponds to $T_0 = T_g - 110K$, which was chosen for analysis and comparison to PEO.

Having examined the trends in both T_g and E_a as a function of r, we can now examine how these inform the observed trend of ionic conductivity with changing r shown in Figure 4.9. Unlike the values for T_g and E_a from r = 0.25 to r = 0.33, σ remains relatively flat. From r = 0.33 to r = 0.5, there is a marked increase in σ of more than one order of magnitude, which continues more gradually thereafter until reaching a maximum at r = 0.75. From r = 0.75 to r = 1, the ionic conductivity declines again. Much like the trends in T_g and E_a, the initial sharp increase in σ

is associated with an increase in r. Unlike T_g and E_a, σ does not demonstrate the same insensitivity to the changes in r. Whereas T_g only varies by a few degrees for $r \geq 0.5$ and E_a only by less than 1 kJ/mol (Figures 4.9B and 4.9A), σ changes more dramatically, increasing by more than a factor of two from r = 0.5 to r = 0.75, and then decreasing by slightly more than a factor of four from r = 0.75 to r = 1.0. It is apparent that in the PBE systems investigated in this work, the conductivity in the salt saturated regime beyond r = 0.33 is driven primarily by the increase in ion concentration and insensitive to both the potential ionic conduction mechanism at play (as indicated by the trend in E_a) and the stiffness of the polymer (as indicated by the trend in T_g). This is a sharp contrast to PEO-based systems, where increases in concentration affect the T_g and thereby the flexibility of polymer chains, which are actively involved in ionic conduction.[38, 51]

It should be noted that in comparing the PBE system in this work to more traditional PEO-based electrolytes, care must be taken to account for differences in experimentally controlled parameters like choice of salt, molar salt ratio, measurement temperature, and the dependent experimental observations like σ and T_g. Research on PEO-based electrolytes has employed a variety of salts, but lithium bis(trifluoromethanesulfonyl)imide (LiTFSI) has become a standard, whereas the choice of salt in this study was a more strongly Lewis-basic lithium tert-butoxide, which was chosen to interact more strongly with the Lewis-acidic boron moieties in the polymer backbone. The molar salt ratios probed in this study and the optimal conductivity observed were also different, with the optimal conductivity of the PBEs measured at $r = 0.75$, whereas PEO-based electrolytes typically demonstrate optimal conductivity at about $r = 0.125$.[17] The PBEs presented here have T_g values ranging from $-16.8°C$ to $59.3°C$, and measurements were taken from 80°C to 30°C in 10°C intervals, a range that includes the T_g of PBE films with $r \geq 0.33$. In a PEO-LiTFSI system, measured T_g values can be much lower than those of the PBE films, with reported values in the literature varying across a range from from $-66°C$ to 5°C.[1] Though Figure 4.9 provides insight into unique characteristics of the PBE systems relative to PEO electrolytes, a robust comparison requires a common reference point.

The VTF equation provides a route to finding a common point of comparison among a variety of polymer electrolyte systems by allowing for a judicious choice of reference temperature, T_0, that can collapse conductivity measurements from different systems onto a common temperature axis by plotting σ against $1000/(T - T_0)$ instead of $1000/T$.[1, 17, 80] By accounting for the glass transition temperature

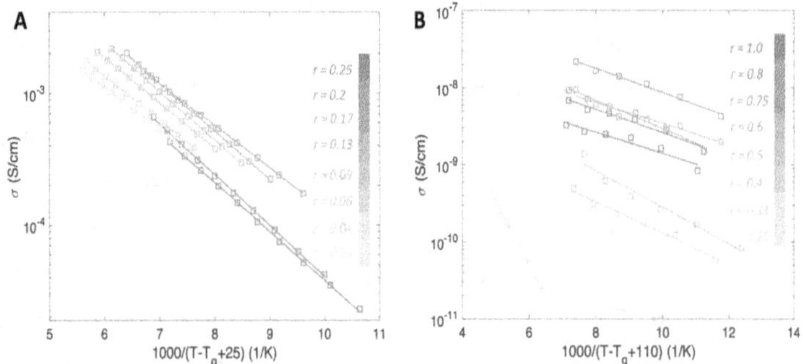

Figure 4.12: Plots of σ against $1000/(T-T_0)$. (A) Plot of PEO-LiTFSI conductivity data from Lascaud et al. [1] against $1000/(T-T_g+25)$. The data for the molar salt fractions shown span a reduced temperature range of 5 to 11. (B) Plot of PBE conductivity data against $1000/(T-T_g+110)$. The data span a reduced temperature range of 4 to 13.

of different systems in this fashion, the role of polymer segmental motion can be controlled for, allowing for the examination of how other parameters like choice of salt and concentration can affect ionic conduction through the consideration of a reduced conductivity, σ_r.[1, 80, 81] We calculated σ_r for both our PBE system and a PEO-LiTFSI electrolyte investigated by Lascaud et al. in a comprehensive study that examined the effect of different salts and their concentrations on the phase behavior and conductivity of PEO electrolytes.[1] To inform the choice of reference temperature, T_0, for the PBE data, we plotted and fit the PEO data from Lascaud et al. (Table 4.3) according to the reference temperature they chose for their system,

$$T_0 = T_g - 25K.$$

Plotting σ against $1000/(T-T_0)$, which upon substituting the choice of T_0, becomes $1000/(T-T_g+25K)$, and results in Figure 4.12A. To collapse the PBE data and fits onto the same reduced temperature axis, we chose $T_0 = T_g - 110K$, which produced the plot in Figure 4.12B. Having collapsed both sets of data onto a common axis, we canmakeachoiceofreducedtemperaturecommontoalmostalltheconc entrations probed in both studies. This implies:

$$PBE: \frac{\overline{}}{T - T_g + 110K} = 8 \tag{4.2}$$

$$PEO: \frac{1000}{1000} = 8 \tag{4.3}$$

$T - T_g + 25K$				
$r = mol_{Li}/mol_{EO}$	$c(mol_{Li}/kg)$	$A(S/cm)$	$E_a(kJ/mol)$	R^2
0.25	2.16	4.245	7.254	0.9994
0.2	1.97	6.434	7.546	0.9994
0.167	1.82	4.696	6.333	0.9998
0.125	1.55	10.08	7.243	0.9998
0.09	1.30	9.516	7.357	0.9998
0.063	1.01	5.165	6.880	0.9998
0.042 0.74	3.354	6.737	0.9999 0.031	0.59	2.327
	6.504	0.9996		

Table 4.3: Activation energy, E_a, pre-exponential factor, A, and R^2 from least squares fit of VTF equation to temperature-dependent conductivity of PEO-LiTFSI electrolyte films at all $r = mol_{Li}/mol_{EO}$ from Lascaud et. al.[1] Li cation molar fraction per mass of polymer, $c = mol_{Li}/kg$, is also shown, which will be used to allow for comparison to PBE systems demonstrated in this work.

for both systems. Solving for T in the equations above results in the following reduced temperatures for each system:

$$T_{r,PBE} = T_g + 15K \tag{4.4}$$

$$T_{r,PEO} = T_g + 105K \tag{4.5}$$

Plugging equations (4.4) and (4.5), in addition to the chosen references for each electrolyte, $T_0 = T_g - 110K$ for the PBEs and $T_g - 25K$ for PEO, into equation (4.1) produced the following reduced conductivities:

$$\sigma_{r,PBE} = A\, T_g + 15\, K^{-1/2} \exp\frac{-E_a}{125 R} \qquad (4.6)$$

$$\sigma_{r,PEO} = A\, T_g + 105\, K^{-1/2} \exp\frac{-E_a}{130 R} \qquad (4.7)$$

The equations above were used to plot the reduced conductivities shown in Figure 4.13, scaled to be shown at the same order of magnitude and plotted against c, molar concentration of lithium per mass of polymer, in order to allow a common concentration scale for comparison.

Having controlled for segmental motion in the polymer electrolyte systems, it is apparent that the PBEs exhibit a similar trend in conductivity as a function of concentration similar to that of PEO. At low concentrations, PBE conductivity is poor,

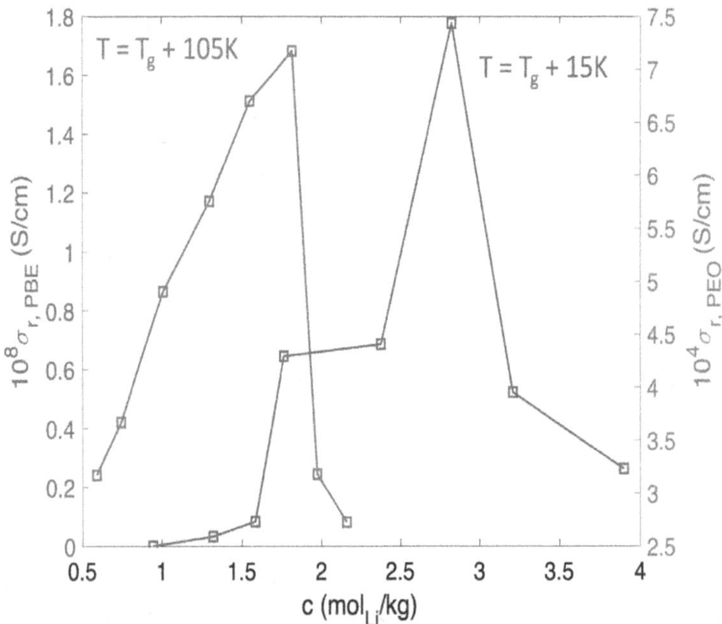

Figure 4.13: Plot of reduced conductivity, σ_r, against Li concentration per polymer mass, c, for both the PEO and PBE systems. Both sets show a similar trend as lithium concentration increases, but the peak for the PBE system occurs at higher salt loadings. but increasing concentration

of ionic carriers leads to a marked improvement until a peak is reached between 2.5 and 3 mol_{Li}/kg, after which the conductivity declines, due to a more complicated solvation environment with pronounced ion-ion pairing and poorer salt dissolution, a trend visible in PEO as well.[1] What distinguishes the PBE system from that of PEO is that this peak in conductivity occurs at ionic concentrations almost two times greater, suggesting that the PBE system can tolerate higher salt loadings and requires these to function. This, in conjunction with the observation that the reduced conductivity of the PBEs is still about four orders of magnitude below that of the PEO system, suggests that the reason for this lies not in the mobility of the chains in the PBEs, but in their chemical makeup (See Figure 4.14) and the overall glassy state of the system. Because the ionic solvation is mediated through the Lewis-acidic boron in the PBE system, a strong base, lithium tert-butoxide, was chosen as the salt. The tert-butoxide anions should interact strongly with the boron centers, dissociating them from the Li cations but binding them tightly to the boron, limiting their mobility. Though Li+ would be relatively free to diffuse compared to the anions, ionic conductivity as typically measured accounts for motion of both anions and cations, and with the anionic motion limited, the conductivity suffers. This is an effect observed in single-ion conducting polymer electrolytes (SIPEs), where Li+ transference is high because the counteranions are bound to the polymer chains, but overall conductivity suffers.[17, 59] This is typically confirmed through measurements of cationic transference number using the procedure outlined by Evans et al.,[82] but the low ionic conductivities presented by the PBEs precluded using this technique to measure Li+ and tert-butoxide anion transference numbers. Furthermore, the 4-coordinate borate centers can become a region of localized negative charge, which strongly attracts the Li+, resulting in poorer salt solvation at higher concentrations and thus lower ionic conductivity. For Li+ to conduct, the concentration must be sufficiently high to decrease the average distance between these boron-anion solvation sites and increase the probability of overcoming the potential barrier associated with this solvation, creating a network of solvation sites accessible only at higher concentrations than what is typically encountered in PEO-based systems.[83] (Figure 4.14).

4.5 Conclusion and future directions

In summary, we developed a novel polyborane electrolyte (PBE) based on a statistical copolymer, poly(9BBN)-co-polybutadiene, which mediated ionic solvation and conduction through the interaction of Lewis acidic boron moieties with strongly Lewis-basic tert-butoxide anions. Characterization via ^{11}B, ^{13}C, and ^{7}Li SSNMR revealed that the poly(9-BBN)-co-polybutadiene was thoroughly incorporated into the polymer matrix, that the UV-crosslinking proceeded to completion, and that the lithium salt, lithium tert-butoxide, was thoroughly incorporated into the electrolyte. ^{11}B SSNMR could not conclusively confirm the interaction of tert-butoxide anions with the boron centers, but EIS experiments on control films without boron revealed that the boron is essential to solvation and conduction of ions in the PBE electrolyte films. Thermal analysis of the films, in conjunction with conductivity measurements and fitting of conductivity data with the VTF equation, revealed that the PBEs demonstrate a relative insensitivity of both the T_g and E_a to changes in concentration beyond

$r \geq 0.5$, while σ showed a more marked dependence on concentration, hinting at a mechanism less reliant on the local viscosity of the polymer and thus the segmental motion of the polymer chains.

Further comparison of the PBEs to PEO-LiTFSI systems studied in the literature

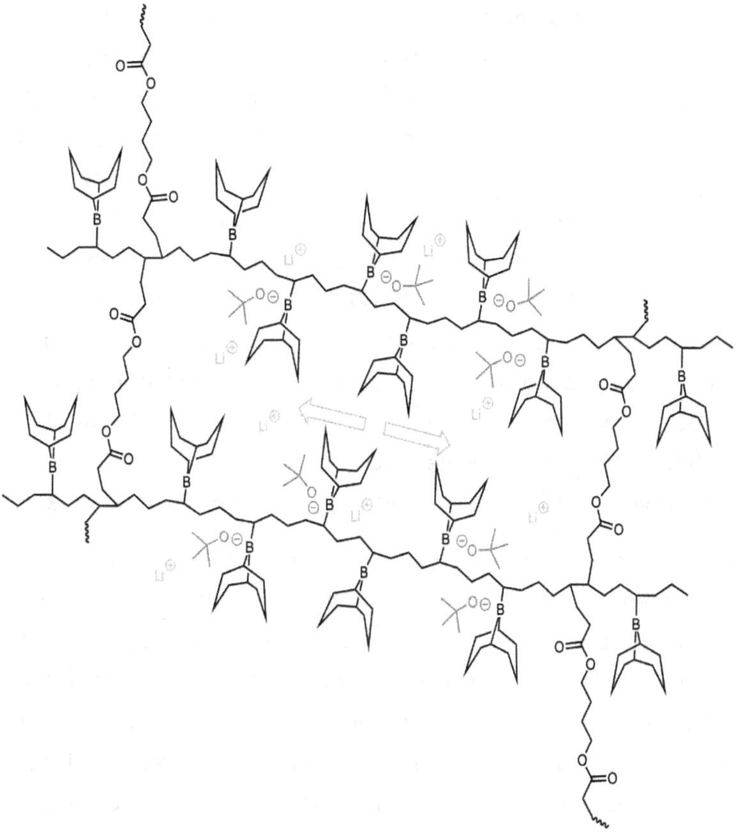

Figure 4.14: An illustration of the structure of the PBE system, where r = 0.5. Note that the tert-butoxide anions are tightly bound to the boron centers. Li+ are attracted to these boron-anion centers, and require them in high enough concentration to maximize the probability of hopping from one site to the next

through the calculation of a reduced conductivity, σ_r, to control for polymer segmental motion revealed the primary importance of salt concentration in driving conductivity in the PBE systems. Consideration of σ_r also revealed that differences between the order of magnitude of conductivity between the PEO and PBE systems, and the concentrations at which they exhibit optimal conductivity, are attributable to the chemistry of the polymer systems, namely the choice of salt and the manner of ionic solvation. The PBEs operate at much lower ionic conductivities than PEO systems because of the presumed strength of interaction between the Lewis-acidic boron sites and the strongly Lewis-basic tert-butoxide anions, in addition to the overall glassy state of the PBE system. This makes the Li+ more mobile relative to the anions, but because ionic conductivity accounts for motion of all ionic species, this ranslates to a poorer conductivity than PEO. The solvation mechanism driven by the interaction of the anions with boron also explains why the PBE systems function optimally at higher concentrations than the PEO-LiTFSI. Li+ ions interact with negatively charged four-coordinate borate complexes, and in order to increase the probability of overcoming the activation barrier to motion associated with this interaction, the concentration must be sufficiently high to decrease the average distance between these interaction sites and provide an accessible network of potential solvation sites.[83]

Further explorations of the mechanism at play could examine the effect of different choices of salt, varying in strength of the Lewis-basic anion, in order to probe how sensitive measured ionic conductivity is to the strength of the boron-anion interaction. It may be the case that softer salts like LiTFSI will result in higher conductivity due to a presumed weaker interaction between boron and the TFSI anions, but it could also be true that the interaction between the boron centers and TFSI anions would not be sufficiently strong to drive dissolution of the salt, a prerequisite for any polymer electrolyte. Investigation of the utility of PBE electrolyte systems would require an understanding of this solvation threshold, and in conjunction with computational study could more precisely parse out the exact mechanism at play in solvation and conduction.

Measurement of polymer network relaxation in PBEs using mechanical characterization techniques like dynamic mechanical analysis or stress relaxation experiments could help elucidate whether the PBEs exhibit super- or sub-ionic conductivity.[84] Superionic solid materials are those that exhibit decoupling of ionic conduction from the motion of the host polymer, and this can be ascertained on a Walden plot of ionic conductivity against characteristic relaxation time where the slope is below 1.[84] Such an investigation could more conclusively confirm whether the PBE system presented in this work achieves some decoupling of ionic motion from that of the host polymer network.

Considering that the PBEs are crosslinked systems, it would not be difficult to incorporate PEO-like elements through addition of UV-curable polyethylene glycol diacrylate(PEGDA)oligomerstotheelectrolyteresins.
Notonlycouldthisprovidea simple route to improving the PBE ionic conductivity, but also a facile means of tuning the solvation and conduction mechanisms at play by varying the ratio of PEGDA to poly(9-BBN)-co-polybutadiene in the precursor resins. In doing so, it might be possible to marry the beneficial characteristics of PEO (higher ionic conductivity) with the anion-solvating effect of the PBEs to improve overall Li-ion mobility, and the higher conductivity would allow for a robust experimental determination of how this combination affects Li+ transference number when compared to PEO systems alone. The ability to cure the electrolyte resin would also allow for the possibility of creating a gel-polymer electrolyte that could demonstrate high conductivity and Li+ transference, and potentially allow for assessment of its performance in a full electrochemical lithium metal cell.

Chapter 5

ADDITIVE MANUFACTURING FOR IR-RESPONSIVE, TIN-COATED MICROSTRUCTURES

5.1 Design of plasmonic materials and how additive manufacturing can expand the design space

When considering the optical properties of candidate materials for use as contrast agents in biological imaging, or as obscurants, electrically conductive materials with demonstrable plasmon resonances in the visible (VIS) and infrared (IR) wavelengths are particularly attractive [32, 33, 85, 86]. Plasmon resonances are a form of non-radiative absorption that arise from the local confinement of electromagnetic oscillations, which is advantageous in the design of particles that optimize extinction efficiency by maximizing absorption [28, 29]. Noble metals like gold (Au) and silver (Ag) are highly conductive, and nanoparticles (nanospheres and nanorods) fabricated from them have demonstrable plasmon resonances in the visual (VIS) and near-IR (NIR) [28, 85–87]. The plasmon resonance of nanorods can be pushed further into the IR by changing their length and aspect ratio [28, 85, 87]. A study by Jain et al. using the Discrete Dipole Approximation (DDA) computed absorption efficiency of gold nanorods with fixed effective radii of 11.43 nm, but increasing aspect ratios from 3.1 to 4.6, which was achieved by increasing their lengths [85]. These calculations revealed that the plasmonic resonance of Au nanorods could be pushed into the NIR, from $\lambda = 727$ nm to $\lambda = 863$ nm [85]. Plasmon resonances can be pushed even further into the IR by increasing these aspect ratios, which could enable the development of new technologies that could improve absorption for solar panels [30], allow direct detection of analytes in biological samples [31], or advance obscurants that function in the near- and mid-IR [32, 33]. Despite their favorable attributes, Au and Ag are untenable as potential IR-active microparticles given their high cost and relative thermal instability [88–91].

Transition metal nitrides like titanium nitride (TiN) offer a promising alternative. Like Au and Ag, they exhibit high electrical conductivity, and nanofilms and nanoparticles fabricated from TiN have demonstrable, tunable plasmonic resonances [89–93]. TiN also exhibits greater thermal stability than gold [88–92], with a thermal conductivity of 15-30 Wm^{-1}K^{-1} at 300 K [94], approximately an order of magnitude less than that of both Au and Ag (314 and 406 Wm^{-1}K^{-1}, respectively) [95]. TiN is also much harder, with a Vickers hardness of 16-18 GPa [94] compared to 188-216 MPa for Au [96] and 250 MPa for Ag [96]. This combination of thermal and mechanical properties makes TiN more suitable for thermally and mechanically demanding plasmonic applications, such as deployment from propellant devices as obscurants [33].

Fabrication of plasmonic nanoparticles is challenging because traditional wet chemistry techniques lack sufficient control over their size distribution, which is important for optimizing extinction [33, 97–99]. Additive manufacturing (AM) processes can enable the fabrication of a variety of structures from diverse materials, with fine control over aspect ratio and feature size [100]. Examples demonstrated in the literature include, but are not limited to: photonic crystals [10, 23, 101]; hollow, 3D, TiN coreshell structures [102]; and 3D graphitic carbon and lithium cobalt oxide electrodes [9, 103].

We developed an AM process that uses two-photon lithography direct laser writing
(TPL DLW) process to create arrays of microbridges with target dimensions of
5 μm (l) × 0.7 μm (w) × 2.3 μm (h) on a sapphire substrate, and we then coat them with a 30 nm-thick layer of TiN using atomic layer deposition (ALD) (Figure 5.1B). Measurements of their optical properties via Fourier-transform infrared spectroscopy (FTIR) reveal broad reflectance attenuation up to wavelengths of 10 μm, and transmittance suppression up to 7 μm. Future advances in the scalability of such AM techniques could enable the mass production of such structures for potential deployment as IR-active microparticles.

5.2 Fabrication of TiN-coated microbridges

The microbridges were first fabricated by printing the polymer scaffolds using a Two-Photon Lithography (TPL) direct laser writing (DLW) process (Photonic Professional, Nanoscribe Gmbh) out of IP-Dip Photoresist on a sapphire substrate. The struts on which the bridges rest were printed using a laser power of 20 mW and a scan speed of 200 μm s^{-1}, and the bridges were printed using a laser power of 30 mW and a scan speed of 200 μm s^{-1}. The structures were then conformally coated using Atomic Layer Deposition (ALD), one monolayer at a time using a FlexALALDSystem(OxfordInstruments)at200 °C.Thedepositionwasperformed by sequentially cycling through the following steps: flowing the reactant dose of tetrakis(dimethylamino)titanium (TDMAT) precursor for 800 ms, purging the

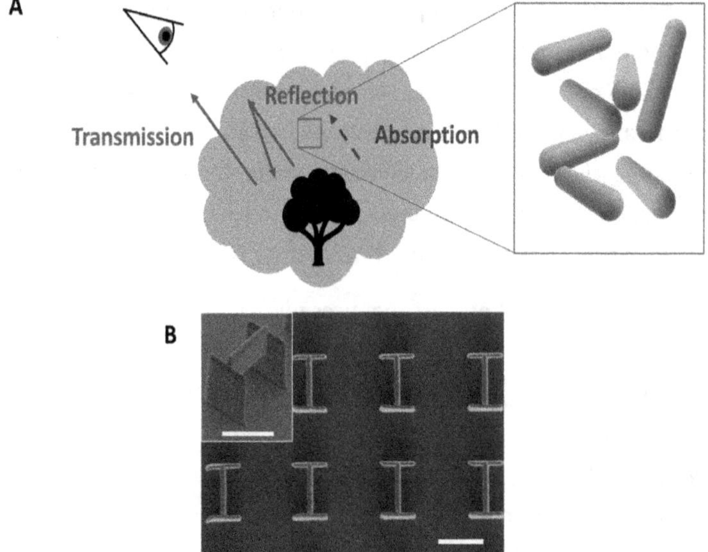

Figure 5.1: (A) Schematic of an IR active material: it should reflect and absorb IR light from potential sources to be distinguishable to an observer. The material is ideally composed of conductive nanoparticles with broad extinction coefficients. (B) Scanning Electron Microscopy (SEM) image of a representative array of 4.749 ± 0.048 μm long, 0.692 ± 0.015 μm wide, and 2.256 ± 0.077 μm high microbridges produced via two-photon lithography and coated with a

30 nm-thick layer of TiN using Atomic Layer Deposition (ALD). Scale bar is 5 μm. The inset contains a zoomed-in SEM image of a typical individual micro-bridge taken at a 52-degree tilt by 45-degree rotation. Scale bar is 4 μm.

system for 3s, plasma treatment with a N2/H2 gas mixture (20 sccm/20 sccm) for 3 s, and purging the system for an additional 2s. An estimated thickness of 0.08 nm TiN per cycle was obtained via ellipsometry measurements carried out on a separate silicon substrate first coated with silica and then with TiN. This process was repeated until an approximately 30-nm-thick layer was deposited.

We examine the morphology of the fabricated samples using Scanning Electron Microscopy (SEM) imaging (Versa 3D DualBeam, Thermo Fisher). SEM images of the finished samples, shown in Figure 5.1B, reveal the microbridges to have a length of 4.749 ± 0.048 μm, a width of 0.692 ± 0.015 μm, and a height of 2.256 ± 0.077 μm, averaged over 8 rods. The microbridges have an effective aspect ratio of 3.368, calculated as a ratio of beam length to the effective radius of a beam with circular cross-section and equivalent cross-section area, which was done to ensure a singular metric for beams with non-uniform cross-section dimensions.

5.3 Material characterization

We characterize the chemical composition of the TiN-coated structures by collecting energy-dispersive X-ray spectra (EDS) on representative samples using the same SEM/FIB system (Versa 3D, Thermo Fisher) equipped with a Bruker Quantax EDS detector. Figure 5.2A contains an EDS spectrum and a compositional map from the top-view of a representative sample. The EDS map reveals a uniform distribution of both Ti and N on the microbridges and substrates, along with no apparent Tior N-rich phases. Quantitative analysis of the EDS spectrum, taken from the node of the rod with one of its supporting struts, reveals a stoichiometric ratio of Ti:N = 0.963, indicative of a slight excess of N but typical for TiN films deposited by ALD [104]. The EDS map in Figure 5.2A also reveals the presence of C, mainly in the rod and its supporting struts, along with some residual C on the substrate. EDS maps and quantitative analysis also revealed the presence of Al in the sapphire substrate, and O, which is also a component of the substrates and present in the polymer scaffold (Figure 5.3).

To further confirm the presence of the TiN-coating on the structures, we collected Raman spectra using a Renishaw M1000 MicroRaman Spectrometer equipped with a 514.5 nm laser (Figure 5.2B). To establish the baseline, we first collected the Raman spectrum from the bare sapphire substrate, which demonstrated the characteristic sapphire peaks at 417, 576, and 751 cm^{-1} [105]. Raman spectra were subsequently taken from uncoated, IP-Dip structures on a sapphire substrate and from the fully TiN-coated structures on a sapphire substrate. The presence of peaks at 150 and 620 cm^{-1} in the spectrum of the TiN-coated structures are attributable to the acoustic and optical modes of TiN, respectively [106–108].

5.4 Optical characterization

To probe the optical response of the TiN-coated structures in the near- and mid-IR, we obtained reflectance and transmittance spectra using a FTIR spectrometer (Nicolete iS50). This instrument is equipped with a Continuum Infrared Microscope (Nicolet) and a rotating microscope stage, which allowed for the acquisition of spectra with the incident polarization either perpendicular or parallel to the longitudinal axis of the bridges. A CaF_2 beam splitter and a white light source were used to access wavelengths from 1.11 μm to 3.33 μm, and a KBr beam splitter in conjunction with an

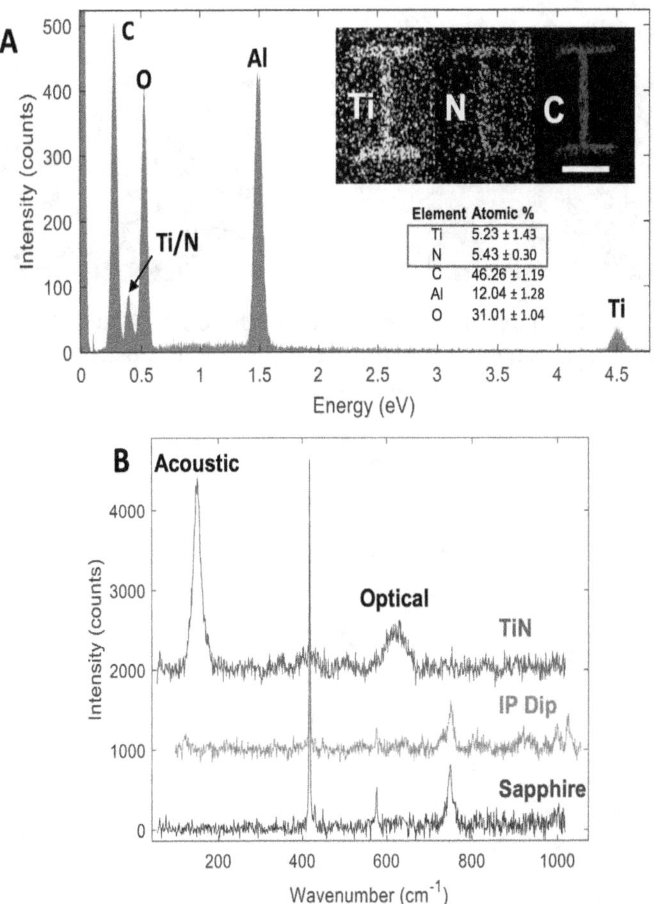

Figure 5.2: Characterization of TiN-coated, microbridges. Image (A) demonstrates the EDS elemental mapping of a single TiN-coated microbridge, along with a representative EDS spectrum with the elemental peaks labeled and a table of the corresponding atomic percentages, with absolute errors included for reference. The scale bar for the mappings is 2 μm. (B) Raman spectra taken from a blank sapphire wafer, IP Dip suspended beams on a sapphire wafer, and TiN-coated microbridges on a sapphire wafer. The presence of TiN is indicated by the labeled acoustic and optical mode peaks.

IR light source were used to access wavelengths beyond 3.33 μm. A CaF₂ polarizer was used for all FTIR spectral acquisitions. Reflectance spectra were obtained over a wavelengthspanof1.11to11.76 μm, andtransmittancespectraoveraspanof1.11to

6.66 μm (Figure 5.6). Transmittance spectra beyond 7 μm were not accessible due to

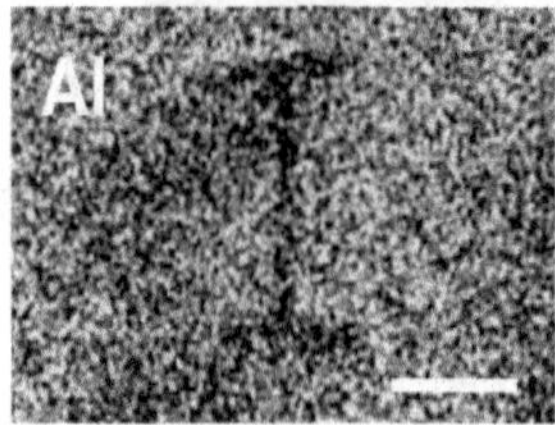

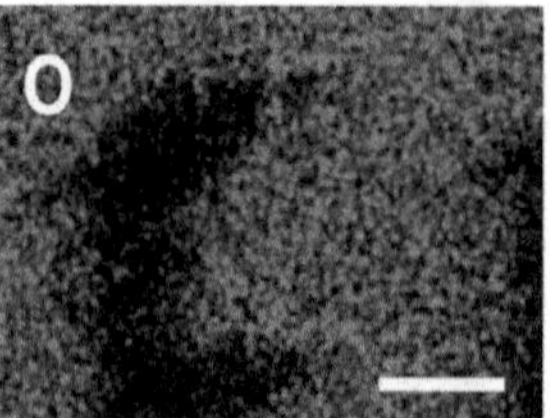

Figure 5.3: EDS mapping of coated structures showing Al and O. The Al comes from the sapphire substrate. O is present in substrate the and the polymer scaffold underneath the TiN coating.

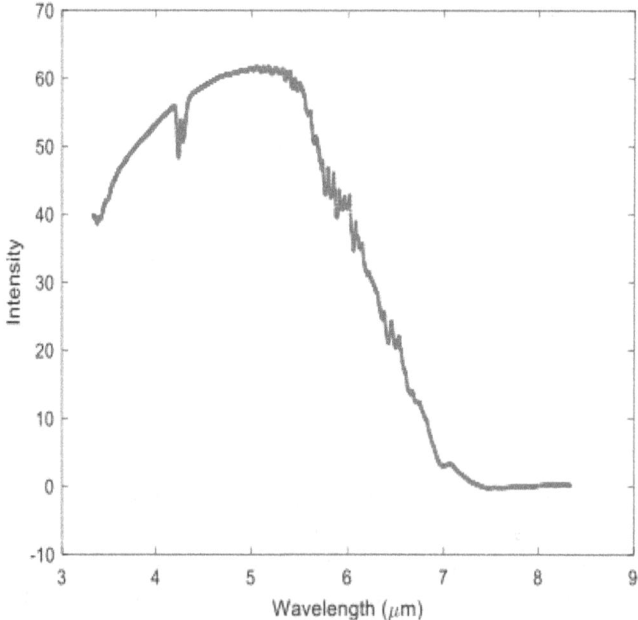

Figure 5.4: Background transmittance signal through sapphire substrate. No light detected beyond 7.5 μm.

the optical properties of the sapphire substrate, which becomes opaque (Figure 5.4). The reflectance and transmittance were subsequently normalized by the background signal obtained from the TiN-coated sapphire substrate. These measurements were also repeated for uncoated, IP-Dip structures on a sapphire substrate as a control to disentangle the effect of the polymer scaffold itself from that of the TiN-coating. Measurements of reflectance from the IP-Dip structures on sapphire between 8.3 to

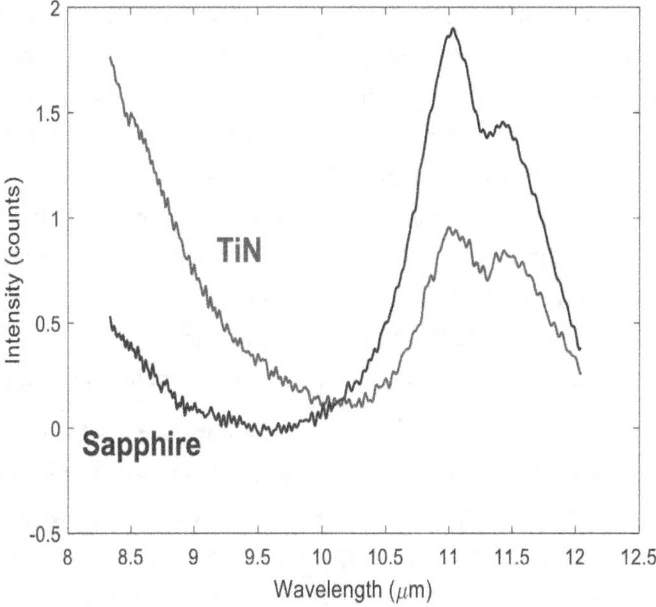

Figure 5.5: Background reflectance signal from sapphire and TiN-coated sapphire substrate. The sapphire signal is effectively 0 between 9 and 10 μm, making the measurement of reflectance in this range for IP Dip control structures on sapphire infeasible.

10.5 μm were not feasible due to the lack of background signal from the sapphire substrate in this region (Figure 5.5).

5.5 Optical response

The reflectance spectra in Figures 5.6A and 5.6B show a pronounced attenuation in reflected light from the TiN-coated structures close to 10 μm. For the case in Figure 5.6A where the polarization of the incident light is parallel to the longitudinal axis of the bridges, the decrease in reflected light is more than 80%, and in the case of Figure 5.6B, when the polarization is perpendicular to the axis of the rods, the attenuation in reflected light approaches 60%. In both reflectance spectra, the TiNcoated structures exhibit greater proportional attenuation of reflected light across the near- and mid-IR than the IP-Dip structures alone on the uncoated sapphire substrate.

The transmittance spectra in Figures 5.6C and 5.6D also display attenuation of transmitted light up to almost 7 μm. When incident light is oriented parallel to the microbridges as in Figure 5.6C, the proportion of light transmitted through them approaches a minimum slightly greater than 65% between 1 and 2 μm, and when incident light is oriented perpendicular to the structures as in Figure 5.6D, the transmittance minimum approaches 65% at about 3 μm. As in the case with the reflectancemeasurements, theTiN-coatedmicrobridgesdisplaygreaterproportional attenuationoftransmittedlightacrossthenear-andmid-IRthantheIPDipstructures alone on the uncoated sapphire substrates.

The peak reflectance attenuation at 10 μm observed in Figures 5.6A and 5.6B also exhibits a dependence on the polarization of incident light. Less than 20% of the incident light is reflected when it is polarized parallel to the longitudinal axis of the microbridges, compared to more than 40% when polarized perpendicularly. This phenomenonisfurtherdemonstratedinFigure5.7, wherethepolarizationofincident light is changed from 0°, to 45°, to 90° relative to the length of the bridges. As the polarization is gradually changed from parallel to perpendicular, the prominence of the dip in observed reflectance decreases incrementally, from less than 20%, to less than 40%, to greater than 40%.

5.6 Discussion

We employed a two-step, additive manufacturing process to create arrays of isolated microbridges with controllable, repeatable feature sizes and optical properties in the near- and mid-IR. We first used TPL DLW to 3D print 4.749 ± 0.048 μm by 0.692 ± 0.015 μm by 2.256 ± 0.077 μm microbridges supported by 5 μm-high supports out ofIPDipresinandthenusedALDtoconformallycoatthesestructureswitha30-nmthicklayerofTiN.Thechemicalandstructuralcharacterizationrevealed thatTiNwas successfully and uniformly coated onto the surface of the structures. EDS analysis revealed an even distribution of Ti and N on the microbridges, and quantitative analysis of a representative EDS spectrum revealed a chemical composition of 5.23 at.% Ti and 5.45 at.% N, resulting in a Ti:N ratio of 0.960, which is within the range expected for TiN. EDS analysis also reveals the presence of C (46.26 at %), which could arise from various sources. C is a substantial component of the IP Dip resin used to 3D print the structures, and EDS maps qualitatively demonstrate greater C signal from the structures than the substrate. The presence of C contamination on the substrate could result from normal handling of the wafers in different laboratory settings where organics can accumulate, such as benches and inside the SEM chamber. The EDS spectrum also reveals the presence of Al (12.04 at%), which arises from the sapphire substrate (Figure 5.3), and O (31.04 at%), which is also a component of the sapphire substrate and apparent in EDS maps (Figure 5.3). O also arises in the structures themselves (Figure 5.3), and can result from the IP Dip used to print them.

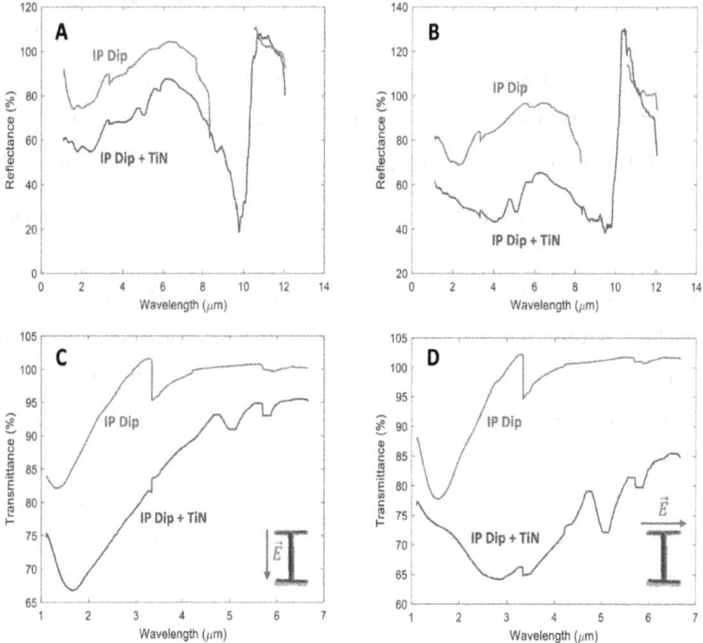

Figure 5.6: Reflectance and transmittance spectra for TiN-coated, suspended rods. The first column shows FTIR spectra when the

polarization of incident light is parallel relative to the longitudinal axis of the suspended rods, and the second column when the polarization is perpendicular to the rods. Images (A) and (B) demonstrate the reflectance TiN-coated structures when relative to that of identical, uncoated, IP-Dip structures. The absence of signal in the reflectance spectra for IP

Dip is because the background signal on sapphire approaches 0 between 9 and 10

μm, making measurement of reflectance from IP-Dip structures infeasible in this range (See Figure 5.4). (C) and (D) demonstrate the transmittance of the TiN-coated structures relative to that of identical, uncoated IP-Dip structures. The TiN-coated structures demonstrate suppressed reflectance and transmittance across the near and mid-IR.

Raman analysis of the TiN microbridges showed peaks at 150 cm^{-1} and 620 cm^{-1} (Figure 5.2B), which are attributed to the transverse acoustic and optical modes, respectively [106–108]. The acoustic vibrations occur in the range of 150-300 cm^{-1} and arise due to the movement of the heavier Ti atoms, and the optical vibrations occur in the range of 400-650 cm^{-1} and result from the movement of the lighter N atoms[107, 109, 110]. TheabsenceofthepeakscharacteristicofTiandNvibrations in the spectra taken from a blank sapphire wafer and from uncoated IP Dip structures on sapphire further confirm the presence of TiN on the coated samples.

FTIR measurements on the TiN-coated, suspended rods reveal broad attenuation of relative reflectance and relative transmittance in the near and mid-IR, when compared to the plain IP Dip structures on an uncoated, sapphire substrate. This signature is preserved for both polarizations of incident light: perpendicular and parallel to the longitudinal axis of the bridges (Figure 5.6). The large attenuation observed in the reflectance spectra in Figures 5.6A and 5.6B at about 10 μm is likely caused by an absorbance resonance in the structures, as incident light is not transmitted through the sapphire substrate at wavelengths beyond 7 μm (Figure 5.4). This dip in reflectance also appears to be dependent on the polarization of incident

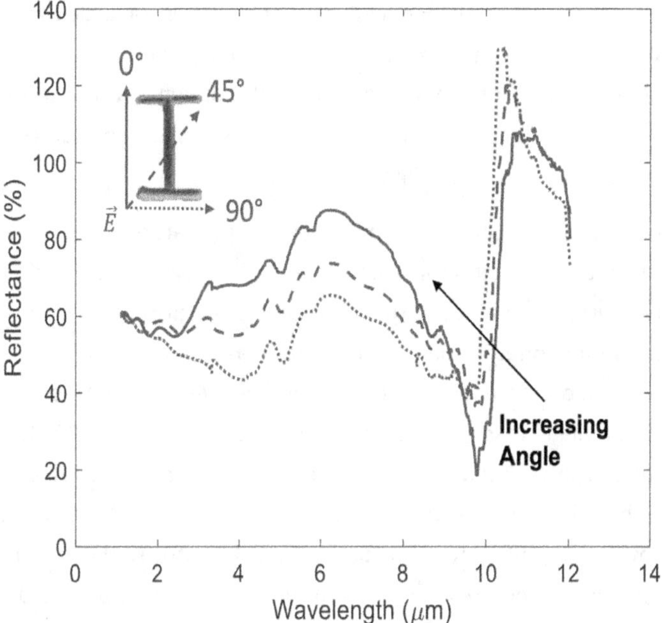

Figure 5.7: Reflectance spectra obtained from the same sample with incident polarization parallel, 45 degrees, and perpendicular to the longitudinal axis of the suspended rods. As the polarization goes from parallel to perpendicular, the presence of the dip between 9.5 and 10.5 μm disappears, and the spectrum blue-shifts.

light. Transmittance appears to be suppressed to a greater degree when the incident light is polarized perpendicular to the longitudinal axis of the suspended rods. The reflectance is at a minimum below 20% when the microbridges are oriented parallel to the polarization of incident light, and is at a maximum of more than 40% when the microbridges are oriented perpendicular to incident light. We also explored this orientational dependance by rotating the structures relative to the polarization of the incident beam (Figure 5.7). As the microbridges are rotated through angles of 0°, 45°, and 90° relative to the polarization of incident light, the dip in reflectance increases from a minimum below 20%, to an intermediate value below 40%, to a maximum above 40%. The reflectance spectra in this range also blue-shift   as   the microbridgesaregraduallybroughtoutofalignmentwiththepolarizationo fincident light. We surmise that the maximum dip in reflectance when the microbridges are aligned with the polarization of incident light indicates a plasmon resonance along the longitudinal axis of the bridges. This is further supported by the attenuation of this feature in the reflectance spectra as the structures are gradually rotated out of alignment with the incident light.

The specific mechanism by which this resonance feature arises remains unclear, but similar structures and phenomena have been studied in the literature. Liu et al. constructed smaller structures from Au where the struts functioned as quadrupole antennas and the bridges on top as simple dipole antennas [111]. The bridges were 355 nm long and exhibited a single resonance in absorbance spectra at ω_b = 170 THz, or λ = 1.8 μm, attributed to dipole-like plasmons in the bridges and observable when they sat along the centers of the struts. As the bridges were moved to off-center positions along the struts, the observed single dipole plasmon resonance from the bridges gave way to a dip in the absorbance spectrum, a result of coupling between the struts underneath and the bridges on top. Our TiN-coated microbridges are also situated along the centers of the supporting struts and exhibit maximum absorbance at around 10 μm when they are aligned with the polarization of incident light, which we infer from the minimum in the reflectance spectra because we know no light is being transmitted through the TiN-coated sapphire substrate. As we rotate the microbridges out of alignment, this absorbance peak also disappears, manifested as a disappearance of the observed reflectance minimum. Though further theoretical study is warranted to precisely understand the mechanism by which this resonance is activated in our structures, we hypothesize that it is plasmonic in nature based on our results and what has been studied in the literature.

5.7 Summary and outlook

In summary, we employ an additive manufacturing process to create TiN-coated microbridges with high effective aspect ratios and tightly controlled feature sizes. EDS analysis confirms the uniform deposition of Ti and N onto the structures and reveals a stoichiometric ratio of Ti:N of roughly 1:1. Raman spectroscopy further confirms the presence of

TiN on the microbridges through the presence of acoustic Ti vibration and optical N vibration peaks in the spectrum. FTIR measurements demonstrate broad attenuation of reflectance and transmittance in the near- and mid-IR, with dependence on the orientation of the microbridges relative to the polarization of incident light. A minimum in reflectance when the structuresarealignedwiththepolarizationofincidentlightsuggestsaprogrammable plasmonic resonance deep in the mid-IR at 10 μm, and further theoretical and computationalstudyshouldelucidatetheprecisemechanismbywhichthis resonance arises. Advancesinimprovingthescalabilityofadditivemanufacturingcouldenable the production of such structures on a large enough scale for meaningful deployment as a future IR-active microparticle of choice.

Chapter 6

CONCLUSION

6.1 Summary and future research directions

This book examined the properties of new materials whose design was made accessible by advances in AM, or driven by the implications of AM materials. In chapters 3 and 4, we designed a novel, poly(9-BBN)-based polymer electrolyte system and characterized its ionic conductivity and glass transition temperature as a function of concentration. Difficulties in obtaining consistently reliable ionic conductivity data from the first generation of these electrolytes described in chapter 3 led to modifications that resulted in the UV-curable second generation covered in chapter 4 that exhibited consistent ionic conductivity as a function of ionic concentration and temperature. Activation energies, E_a, ob tained fr om VT F fi ts of th e conductivity data, in conjunction with glass transition temperatures measured from DSC, potentially reveal a unique ionic conduction mechanism where lithium ions are free to move more independently of polymer segmental motion at high salt concentrations along anion-boron complexes in the polymer backbone. Further computational study could could more precisely elucidate the ionic conduction mechanism, and investigation of their mechanical relaxation times via techniques like step-strain relaxation experiments could help clarify whether these materials are sub- or superionic conductors. Investigation of a gel version of these electrolytes could increase their ionic conductivity sufficiently for actual application in 3D batteries.

Chapter 5 examined the optical response of TiN-coated microbridges manufactured via a TPL DLW process, follwed by ALD. Characterization of the optical response of these structures in the near- and mid-IR revealed broad attenuation of reflectance and transmittance, attributable to the TiN coating and the shape of the structures. Moreover, a pronounced dip in the amount of reflected light at wavelengths of close to 10 μm, w hich c hanges d epending o n t he orientation of the microbridges relative to the polarization of incident light, suggests a plasmonic resonance deep in the IR. Further computational study should clarify the precise mechanism at play, and future investigations can explore how factors like array spacing and the length of the bridges can affect the observed dip in reflectance.

6.2 Outlook

The work presented in this book illustrates how two distinct, and at times seemingly unrelated projects are united by the common thread of AM. In the case of the polymer electrolytes presented in chapters 3 and 4, their design and characterization were motivated by the push for better batteries that can store more energy without sacrificing power, which is essential in the push for more sustainable modes of transportation and energy generation. AM has allowed for a complete reconsideration of typical battery design, where the typical thin film stack of anode, electrolyte, and cathode can be replaced by architected structures that in theory would allow us to circumvent the energy/power tradeoff inherent in current lithium-ion batteries. These batteries are not possible, however, without a solid electrolyte that has the requisite ionic conductivity and can be conformally coated onto the complex, 3D electrode geometries that AM can produce. This led us to consider polymer electrolytes, more specifically a novel poly(9-BBN) based electrolyte, that allowed us to explore a potentially new mechanism of ionic solvation and conduction. Furthermore, refinement of the first generation of this electrolyte led to a UV-curable second generation, which could find application as a conformally coatable electrolyte membrane for architected batteries. It is my hope that further investigation of the mechanism of ion conduction at play in these polymer electrolytes can provide further insight into improving this class of materials as a whole, thereby enabling the next-generation energy storage technologies needed to meet the growing climate challenge.

The TiN-coated microbridges presented in chapter 5 demonstrate how AM technology directly expands the space of materials and material properties available for exploration to materials scientists. Plasmonic materials have been studied primarily in the visible and near-infrared, owing to the materials traditionally studied—like Au and Ag—and the standard chemistry techniques employed to make plasmonic nanoparticles. AM affords us the means to create particles of different shapes and sizes uniformly, and at different lengthscales, depending on the AM process employed. This has the potential to push the study of plasmonic nano- and mircoparticles further into the IR, which we explored by using TPL DLW to write our microbridges and then coated in TiN using ALD. The ability to create such optically active structures deep into the IR has implications for a wide-variety of applications, from solar panels that can absorb more efficiently over a broader range of wavelengths, to materials that can be employed as IR obscurants. With future advancements in improving the scalability of the AM processes, along with further investigation of the exact mechanism responsible for the appreciable attenuation of reflected IR light from our structures, it is hoped that microparticles like ours can become more accessible for a variety of IR applications.